IMAGES
of America

ELEPHANT BUTTE DAM

ON THE COVER: This photograph of Elephant Butte Dam shows water being released from one of the dam's gates. The dam was part of the Rio Grande Project, and the water would be used for the irrigation of the Rio Grande Valley. (Courtesy of Sherry Fletcher.)

IMAGES
of America

ELEPHANT BUTTE DAM

Sherry Fletcher and Cindy Carpenter

ARCADIA
PUBLISHING

ISBN 978-1-4671-3332-6

Published by Arcadia Publishing
Charleston, South Carolina

Printed in the United States of America

Library of Congress Control Number: 2015937264

For all general information, please contact Arcadia Publishing:
Telephone 843-853-2070
Fax 843-853-0044
E-mail sales@arcadiapublishing.com
For customer service and orders:
Toll-Free 1-888-313-2665

Visit us on the Internet at www.arcadiapublishing.com

To Baxter, John, Maxine, Jessica, Eve, Vicki, Penni, Amanda, and Milo Ty

—All my love, Sherry

To Jay, Beau, Amanda, Jake, Racheal, Jess, Cindy Lou, Imogene, Sherry, Jamie, Kim, Lee, Cordell, Colten, Amazing Grace, and Evie Lou

—I love you all dearly, Cindy (Nana)

Contents

ACKNOWLEDGMENTS

We would not have been able to complete this book without the editing assistance of Ann Burris-Welborn, Cindy Lou Burroughs-Carpenter, Dr. Gabrielle Palmer, Russell and Melissa Woolf, and Michelle Losey. We were so pleased and honored to receive historical photographs from Chuck, Karen, and Brittany Martin, Carol (Woolford) Kipp, Phil and Bobbi Woolford, Vicki Cecil-Sears, and the estates of Annette Wilson-Smith and Juanita Mims-Nelson. A special thanks goes to Moshe Koenick, who was instrumental in locating the Laverne Charles Collection of dam photographs that Sherry was able to purchase. All photographs, unless otherwise credited, are from the Laverne Charles Collection and have been copyrighted by Sherry Fletcher. Also, we would like to thank Moshe for allowing us to use his rare early dance invitation from an Elephant Butte Ball. Kay Dunlap, representing the New Mexico State Parks, was instrumental in securing the photograph of the CCC camp and history. Thank you Kay for your assistance. Tiffany Berger, soon to be Dr. Berger, selflessly gave of her time to locate information and research. Thank you Tiffany for being so generous and helpful. A huge debt of gratitude to Penni Lane Sears for all of the scanning and organizing of photographs from the Laverne Charles Collection. We value the love and support of our families during this time of messy tables, papers scattered everywhere, and minds adrift. We tried our best to multitask but have decided that it is totally impossible for two old ladies to do so. Therefore, we appreciate your love, support, and belief that we can change the world through history. And last, but not least, we want to thank our amazing editor at Arcadia Publishing, Stacia Bannerman. She was supportive and positive through both Images of America: *Hatch Valley* and this book. Thank you, Stacia.

Introduction

Early residents of the New Mexico Territory depended upon manmade ditches called *acequias* as an irrigation system to divert water from the Rio Grande to their farms, animals, and villages. As population increased in the western territory and naturally occurring droughts continued to plague the arid land, the demand for the precious resource from the Rio Grande increased. As the water usage increased to the north, the Mexican farmers pressured their government to intervene for their part of the water. As the water to Mexico continued to dry up, the Mexican government filed a claim of $20 million in damages against the US government on the grounds that Mexican farmers had been deprived water because of a large number of ditches drawing water from the upper portion of the Rio Grande.

Determined to come up with their own solution, Mesilla Valley residents in southern New Mexico formed the Rio Grande Dam and Irrigation Company in 1893. They proposed building a dam across the Rio Grande at a point called Elephant Butte near Engle, thus establishing a reservoir they would call Lake Esperanza. The Rio Grande Dam and Irrigation Company claimed they would give rights of appropriation of water for ditches and canals to Doña Ana County and Sierra County. Conspicuously, Mexico was left out of the deal. The secretary of the US Department of the Interior approved the idea with the caveat that the project had to be completed within five years.

Through English investors, Dr. Nathan E. Boyd was able to obtain the necessary financial backing for the company. Boyd himself held a large number of shares of capital stock in the company. In 1896, boundary commissioner general Anson Mills came up with what he considered a better alternative to the Elephant Butte project. He proposed an international dam be built at a site three to four miles north of El Paso, situated on the Mexico–United States border, at a cost of around $300,000. Anson felt that this would establish a reservoir to stabilize the flow of the river and better serve the irrigators below. With mounting pressure from Mexico to go along with the international dam proposed by Mills, acting attorney general Holmes Conrad filed an injunction against the Rio Grande Dam and Irrigation Company in May 1897 to stop the Elephant Butte project.

Residents north of El Paso were not in favor of the international dam and the project was characterized as the "Anson Mills' Scheme." On May 21, 1897, the *Rio Grande Republican* editorialized that there were those who wished to "induce Congress to make a fish and duck pond for the benefit of El Paso hotels, restaurants, and lodging houses. If it were irrigation they were after, the Elephant Butte site is far more eligible to furnish water for the El Paso Valley than the international job."

According to W.B. Childers, US district attorney for New Mexico, the Rio Grande was a navigable stream up to El Paso and that building the dam at Elephant Butte would be illegal based upon the River and Harbor Act of 1890, which prohibited the creation of any obstruction not affirmatively authorized by law. He alleged that the Rio Grande Dam and Irrigation Company would be building a dam to use all the water of the Rio Grande, preventing the natural flow of the Rio Grande and obstructing the navigation of the river below.

Mesilla Valley residents who were backing the Rio Grande Dam and Irrigation Company claimed Congress had sent a special commission to the area that had reported the Rio Grande was not and could not be made navigable for at least 200 miles below El Paso. They argued that the Mexican government and others in Congress were now saying the river was navigable in an attempt to stop the Elephant Butte project. A petition with 300 signatures was signed by residents in the Mesilla Valley against the Mexican and US governments' claim that the river was navigable. They sarcastically remarked that if US Attorney General McKenna had passed through El Paso when the Rio Grande was at flood stage, then he might have thought the Rio Grande was navigable. The residents also feared that building the international dam would lead to the Mesilla Valley becoming the reservoir for the dam, destroying their fertile valley.

The Third District Territorial Court Judge Parker would hear the case in Las Cruces, New Mexico. The US Supreme Court had already ruled the government had control over all navigable streams to their sources. Judge Parker would have to decide if the proposed dam at Elephant Butte would impair the navigability of the Rio Grande.

Witnesses on both sides were called in to testify during the proceedings. Judge Albert B. Fall was called in on December 29, 1899, to continue the argument for the defense. On January 2, 1900, the Frederick, Maryland, newspaper, the *News*, ran an article from El Paso, Texas. In the article, it said District Attorney Childers had contacted Col. Warren Sutton, associate counsel for the government in the Elephant Butte Dam case, and notified him Judge Parker had ruled against the government. The English company could go ahead and build its dam at Elephant Butte for the purpose of supplying irrigation water to American and Mexican territories. Refusing to give up, Colonel Sutton warned that the government would take the case all the way to the territorial and US Supreme Courts. His threats, however, were for naught, as the Rio Grande Dam and Irrigation Company was found to have forfeited its rights to build the dam by the Territorial Court of New Mexico on May 3, 1903. The project had not been completed in the allocated five years.

The government, having no solution, still had to contend with the Mexican government's claim of $20,000,000. Discussion about building an international dam continued. The government of Mexico, feeling the root of the problem was the United States, believed the Americans should bear all costs of the international dam. The United States agreed to the actual terms but did not like the site, fearing the El Paso dam would cause flooding and would be too far downstream to even begin to solve the overarching problem. The El Paso side wanted the international dam to be put across the river at Courchesne, Texas. Other sites, including an area near Albuquerque and the present-day Elephant Butte area were then proposed. For several years, the three parties, New Mexico, El Paso (Texas), and the Republic of Mexico waged verbal warfare over damsites.

On November 18, 1904, engineer B.M. Hall of the US Bureau of Reclamation submitted his final report to the Twelfth National Irrigation Congress in El Paso, Texas. His concluded the site at Elephant Butte would be the best solution because the mountain slopes were a natural basin for the new lake.

A year after the agreement was made in El Paso by the Twelfth National Irrigation Congress, plans were drawn up for the dam. The responsibility for the building of the dam lay with three American civil engineers: Arthur P. Davis, director and chief engineer for the US Bureau of Reclamation, Louis C. Hill, design engineer, and E.H. Baldwin, construction engineer. A Congressional Act on February 25, 1905, authorized the provision for the construction of the Engel (Engle) Dam, named for the nearest train stop on the railroad at Engle.

The December 15, 1906 issue of the *Galveston Daily News* reported that the Mexican Government had ratified the treaty with the United States, providing for the dam to be built and allowing equitable distribution of water. Mexico would withdraw its original $20,000,000 claim against the United States. In 1907, the General Treasury of the United States appropriated $1 million non-reimbursable funds toward the construction of the Engle Dam. Under the Treaty of 1906 with Mexico, 60,000 acres of water was to be delivered annually.

Fearing that private contractors would only cause delays, the Bureau of Reclamation Service decided to use government workers. Construction began on September 9, 1911, and 150 men were used to blast the roadbed and begin construction on the bridge with the first "shot" of 2,000 pounds of powder. The headline in the *Albuquerque Journal* read, "Enormous 'Coyote Shot' Shakes Valley Around Project." The 10-mile-long railroad spur that would be the link to the Santa Fe Railway between Engle and Cutter was also built that same year.

An appropriation of $6,355,000 was given in 1911 to help with the dam's construction costs and the cost of the irrigation structures that would be in place downstream. Two diamond drills were excavating the foundation of the dam by January 11, 1911. An average of 850 workers would be employed by the government to construct the dam over the next five years. The Bureau of Reclamation began filling the reservoir in 1915. The last light post signaling the end of the construction and dedication would take place October 19, 1916.

Pres. Woodrow Wilson did not attend but sent A.A. Jones as his representative, along with Brig. Gen. George Bell Jr. and Brig. Gen. G. Morton, commanders of the 10th and 11th Divisions of the US Army. A company and band from the 23rd United States Infantry escorted Jones as he arrived from Cutter by special train. There were 350 delegates of the International Irrigation Congress and International Farm Congress, along with other guests, lining the banks of the Elephant Butte Lake and standing on the top of the concrete dam. Jones stood on a platform built over the spillway of the dam, shadowed by the elephant-shaped rock structure in the background. He delivered his dedicatory address on behalf of the president to the crowd.

"I am told," said Jones, "this dam at the top of is 16 feet long, that the base is 205 feet thick, and is anchored to a foundation 318 feet below the crest; 610,000 cubic yards of concrete are bound together by interlacing bars of steel. These million tons of material is sufficient to construct on a city lot 25 feet by 125 feet a pillar of concrete lacking on 50 feet of being a mile in height. In its bosom are fixed attachments for the generation of the equivalent of 3,500 horsepower in electrical energy. The reservoir in storage capacity surpasses any other ever constructed. When filled, it will contain more than 2,600,000 acre-feet of water—enough to cover a depth of one foot an acre of 4,200 square miles. This is two-thirds more than the combined capacity of all the reservoirs built or projected for the city of Greater New York. It will actually submerge more than 42,000 acres of land, and extend to a maximum length of 45 miles. The project is designed to irrigate and enable the intensive cultivation of 180,000 acres of land."

From Engle Dam to Elephant Butte Dam, there were other names placed upon the dam during its construction. On December 10, 1912, Eugene Manlove Rhodes (a famous New Mexico writer who was nicknamed the "Cowboy Chronicler") wrote a letter to the Reclamation Service complaining that the name of Engle Dam had been changed to Eagle Dam according to Eastern newspapers. Rhodes was also upset with the Santa Fe Railroad for changing the name of the railroad station in Engle to Engel in 1920. It had originally been called Engle to honor a construction engineer named R.L. Engle. In 1920, it was changed to Engel in honor of Santa Fe Railroad vice president Edward Engel. Rhodes and others complained to Congress, and the post office kept the original spelling of Engle. The railroad was not so conciliatory and left everything associated with the railroad spelled Engel. No wonder the newspapers were so confused. It had to have pushed Rhodes over the top when the *Rio Grande Republic* of October 20, 1916, wrote an article that said the dam had been "Christened the Woodrow Wilson Dam." The story said, "We are met to dedicate to the use of the present and future generations this massive structure and the impounded waters, which, in accordance with the action of the Water User's Associations of Las Cruces and El Paso, shall henceforth bear the name of the President of the United States—Woodrow Wilson."

The Elephant Butte Reservoir had its share of name changing also. Many of the early newspapers called it Lake Hall after engineer B.M. Hall of the US Bureau of Reclamation. That was not the only name change proposed over the years.

Two buildings at the damsite that had been used as barracks and mess halls by the camp workers were auctioned off by the government. The buildings were dismantled and hauled in pieces by wagon to the town of Palomas Hot Springs. There, they were reconstructed and would become the Vera Hotel and James Apartments.

Many of the territorial newspapers quoted statistics on the dam during the construction that may or may not have proved true. In the end, the gravity dam was constructed with 608,000 cubic yards of masonry at a cost of $5 million. It had a length of 1,310 feet and a width of 215 feet at the base. The top of the dam had an 18-foot roadway. The height of the dam was 306 feet above the foundation and 205 feet above the river. Elephant Butte Lake was 45 miles from north to south and had 250 miles of shoreline and 31,000 acres of water surface. The greatest depth was 193 feet, and the average depth was 66 feet.

No, there was never any evidence of any bodies being buried in the cement during the building of the dam. That would have significantly compromised the structure of the great dam. Elephant Butte Reservoir would become part of the largest state park in New Mexico, which became known as Elephant Butte Lake State Park in 1964. The dam was closed to traffic after the September 11, 2001, terrorist attacks on US soil, but it has been opened to the public on special events. In 2016, the dam will celebrate its centennial. Happy birthday, Elephant Butte Dam.

One

Preparing the Land

The Damsite

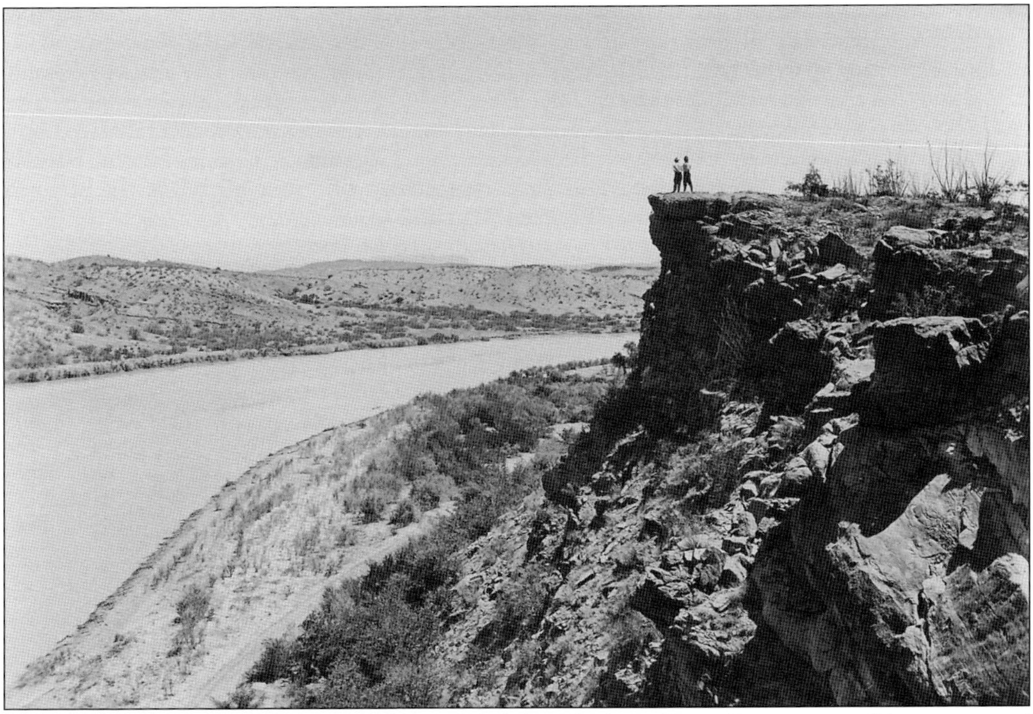

In May 1905, the *Rio Grande Republican* called the Elephant Butte Dam project "the greatest of its kind in the world with the single exception of the Great Nile Dam in Egypt." The photograph above shows the Rio Grande before the construction of the diversion channel that would have to be built to allow the Rio Grande to continue its flow down the valley during the massive project.

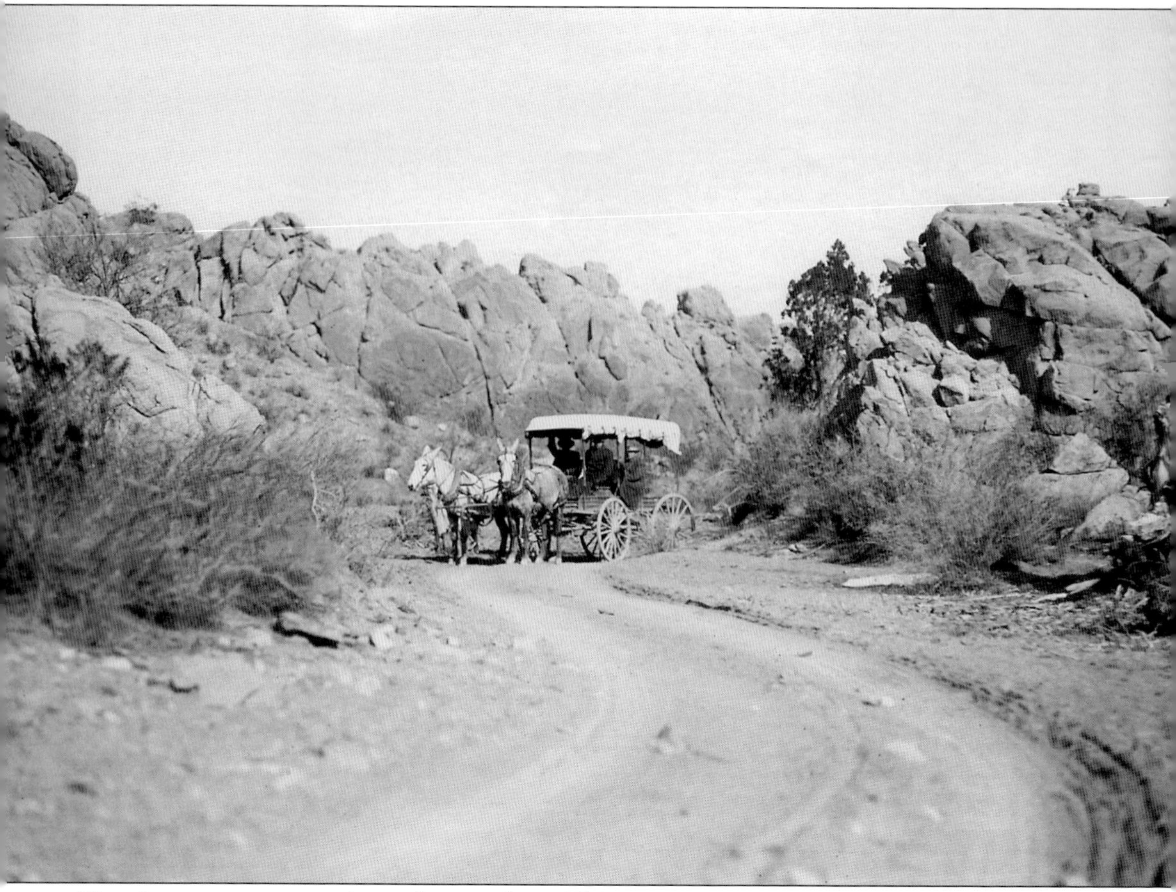

Traveling in a wagon similar to the one above, El Paso attorney P.W. Dent, in connection with the Reclamation Service, spent three weeks securing 37 purchase contracts from Mexican ranchers, and "he wouldn't stop" it was said until he had "every foot of the 40,000" acres secured for the planned Engle (later to be changed to Elephant Butte) Reservoir. The purchase price, according to value, went from $5 to $23 an acre. Dent was accompanied by a native interpreter and was noted to have met only one American, Fred Hanna, who had the store at Canta Recio. He had noted to the paper that he had found the people in the area to be very primitive, living much as they had a century or two ago. He was amazed that the people still used flint and steel for starting fires and lighting cigarettes instead of the more modern-day matchstick. He was buying up the property from individuals living in the ancient villages of Paraje, Canta Recio, Cantadero, San Albino, San José, and Alamosita.

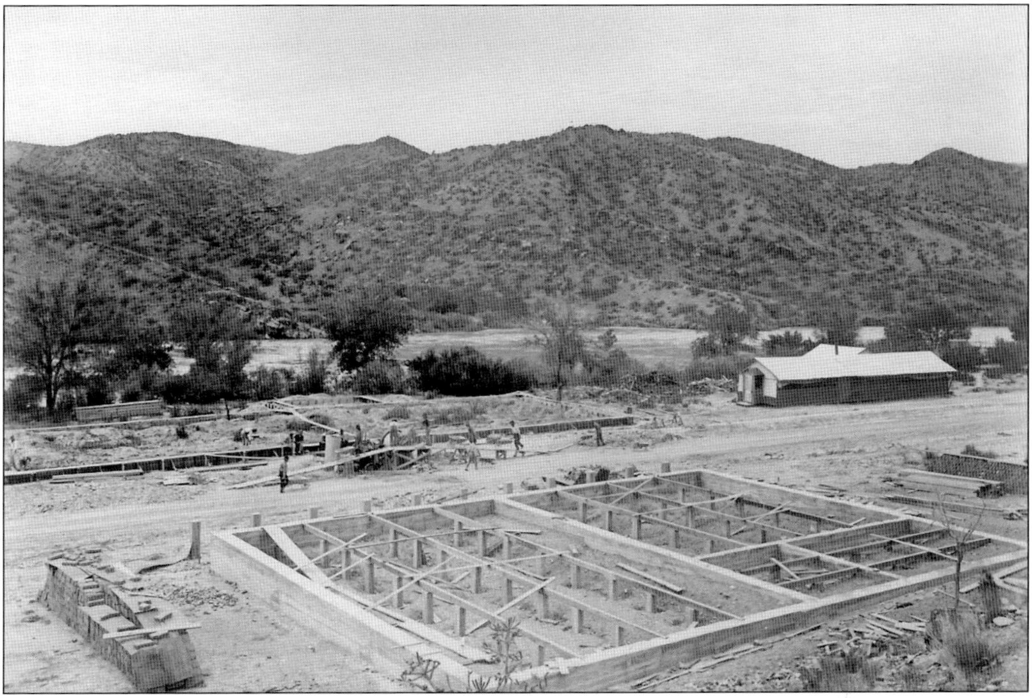

Victorio Land and Cattle Company of California owned three-fourths of the land the new dam and reservoir would occupy. They wanted $17.83 an acre, but the US Reclamation Service offered $1.83 an acre. While they argued in court in the spring of 1909, a group of 150 men were busy erecting facilities at the damsite. A final settlement of $6.66 an acre for the 30,000 acres was finally reached.

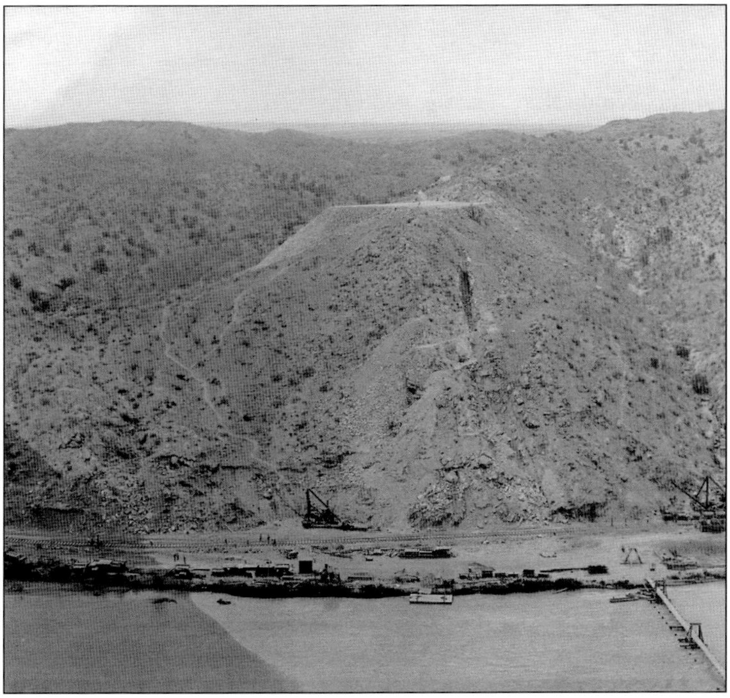

This early photograph shows the damsite in June 1911 looking west toward the spillway. As early as November 26, 1909, the *Rio Grande Republican* had reported on the preparations being done: "We will bring the machinery from the Roosevelt and Yuma damsites and buy the additional heavy machinery necessary for the work at Engel." At that time, the dam was known as the Engle (Engel) Dam.

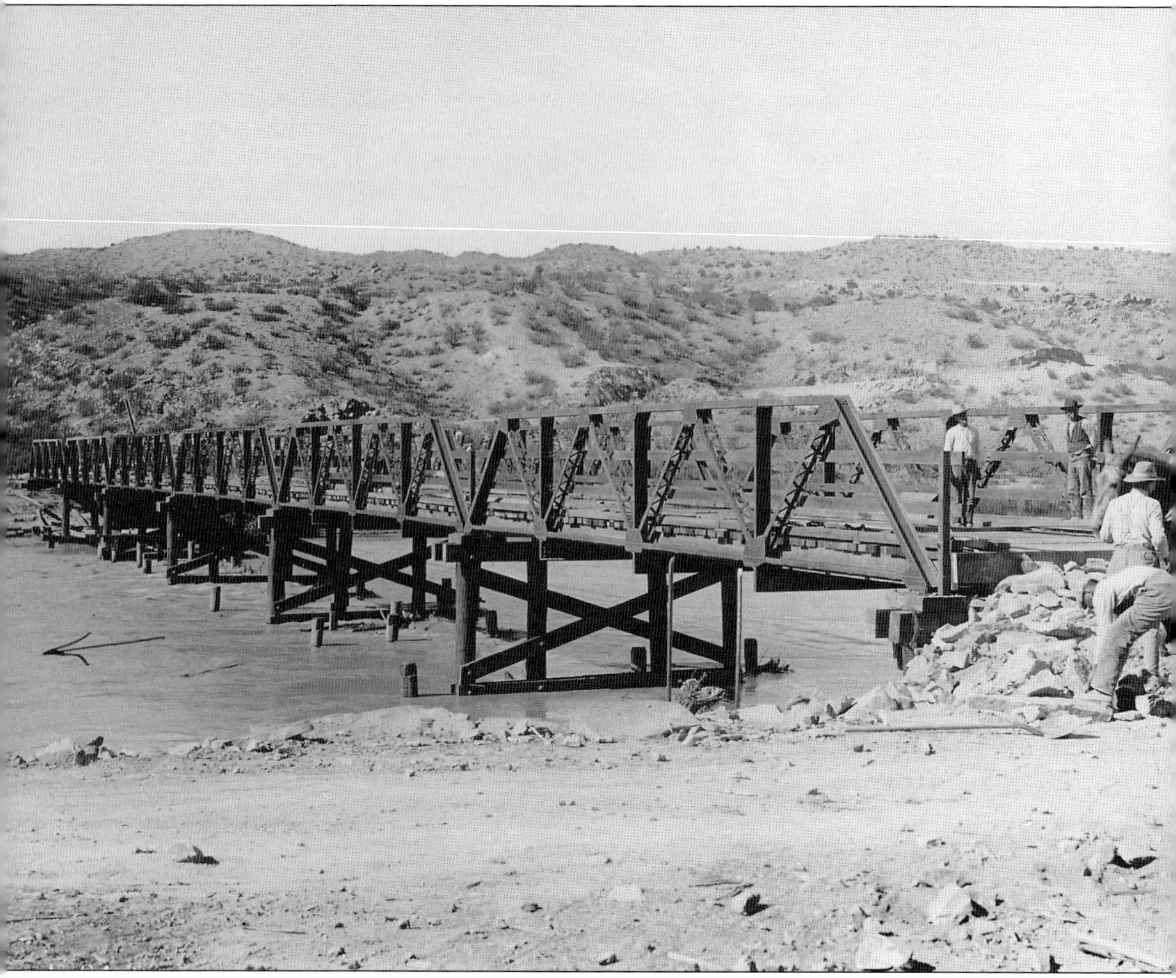

The Reclamation Service sent out advertisements for bids for the construction of a wagon bridge to be built across the Rio Grande. The bridge would be built a short distance down from the actual damsite and would allow for wagons to travel to and from the damsite with supplies and equipment. It was said that the bridge would also be used by the local farmers and others traveling in the area. The photograph above shows the actual work being done on the bridge in 1911. That same year, a contract had been given to a Mr. Oliver of Cutter, New Mexico, for the making of 80,000 adobe blocks to be used in the construction of the office building, cottages, and other buildings at the damsite. The government had already purchased 20,000 blocks, and the additional 80,000 adobe blocks would allow for the construction to continue.

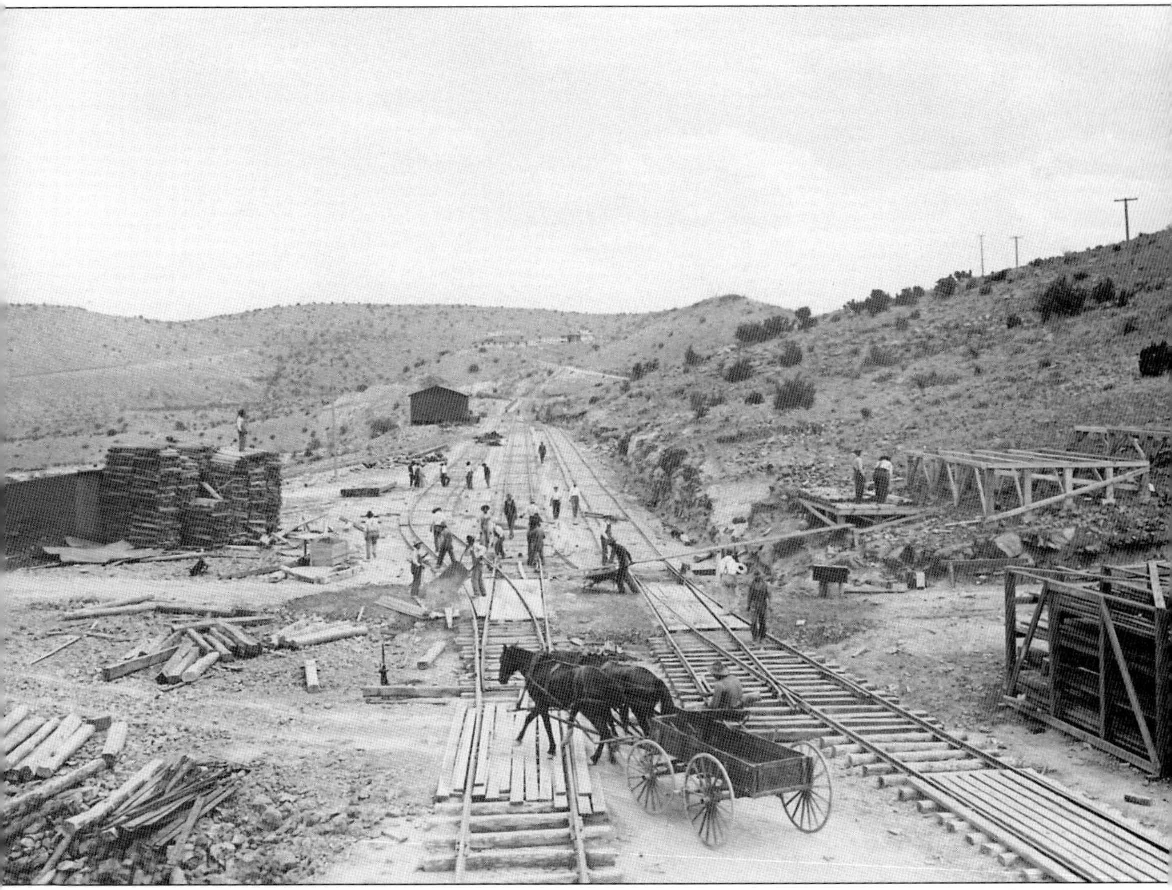

By March 1911, the name of Elephant Butte had been given to the post office established at the damsite. Although the branch railroad was not quite complete, it was at a point where materials could go by rail rather than having to come the 12 miles from Engle by wagon. There were roughly 300 men now employed at the damsite. Two railroad cars of sewer pipe had been received and would be used in the construction of the sewer system for the camp. The Reclamation Service had purchased 50 tons of alfalfa. One carload was at Engle in February 1911 and would be hauled to the camp site by wagon. According to the *Rio Grande Republican* from February 10, 1911, "A considerable quantity of supplies and materials of all kinds is being ordered to be delivered as soon as the branch railway has been completed. A carload of corrugated roofing is expected to arrive in a few days and will be used in the construction of warehouses, etc."

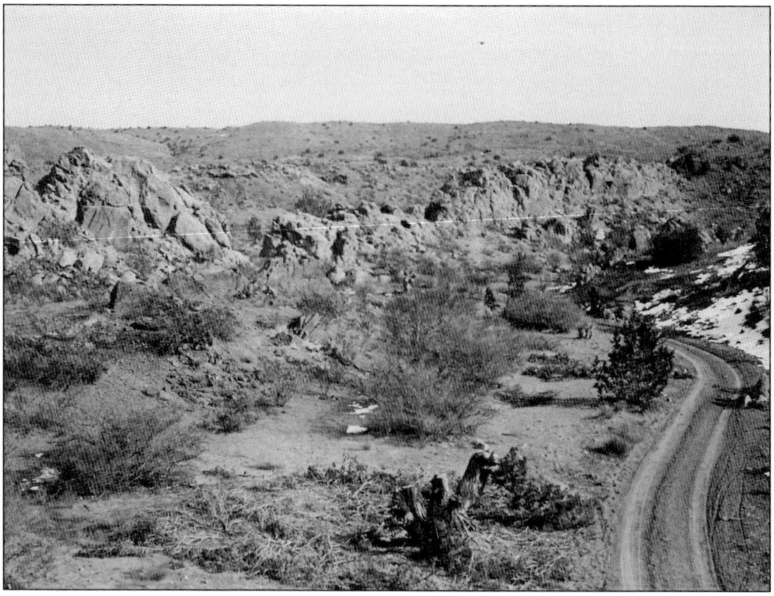

The *Rio Grande Republican*'s headline of June 10, 1910, said it all: "Elephant Butte Dam A Ponderous Undertaking." The article called the preparation work gigantic. One of the preparations that had to be done was the grading of the roads around the damsite in order to prepare for the massive amount of freight that would be hauled over roads and by rail.

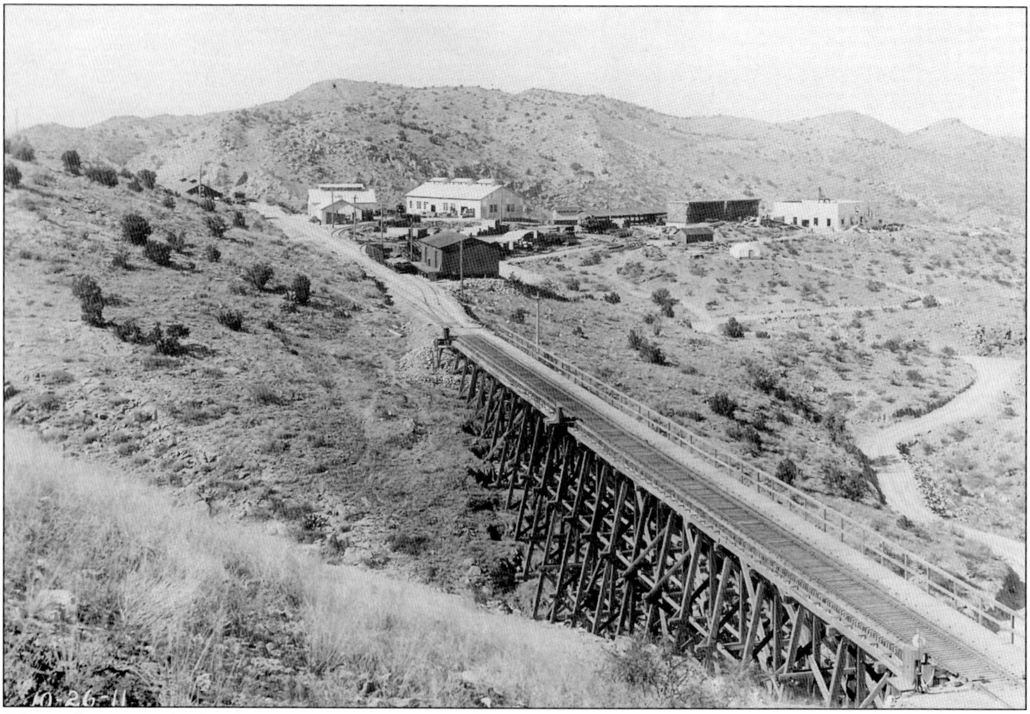

This photograph, looking northwest toward the Rio Grande, shows the railroad yards, shops, and the power plant. The *Albuquerque Journal* on November 21, 1911, reported, "The contract work of W.E. Anderson of El Paso, contracting engineer, will be completed on Saturday night of this week." The power plant was only to be used in the construction of the dam and would be dismantled at the completion of the dam.

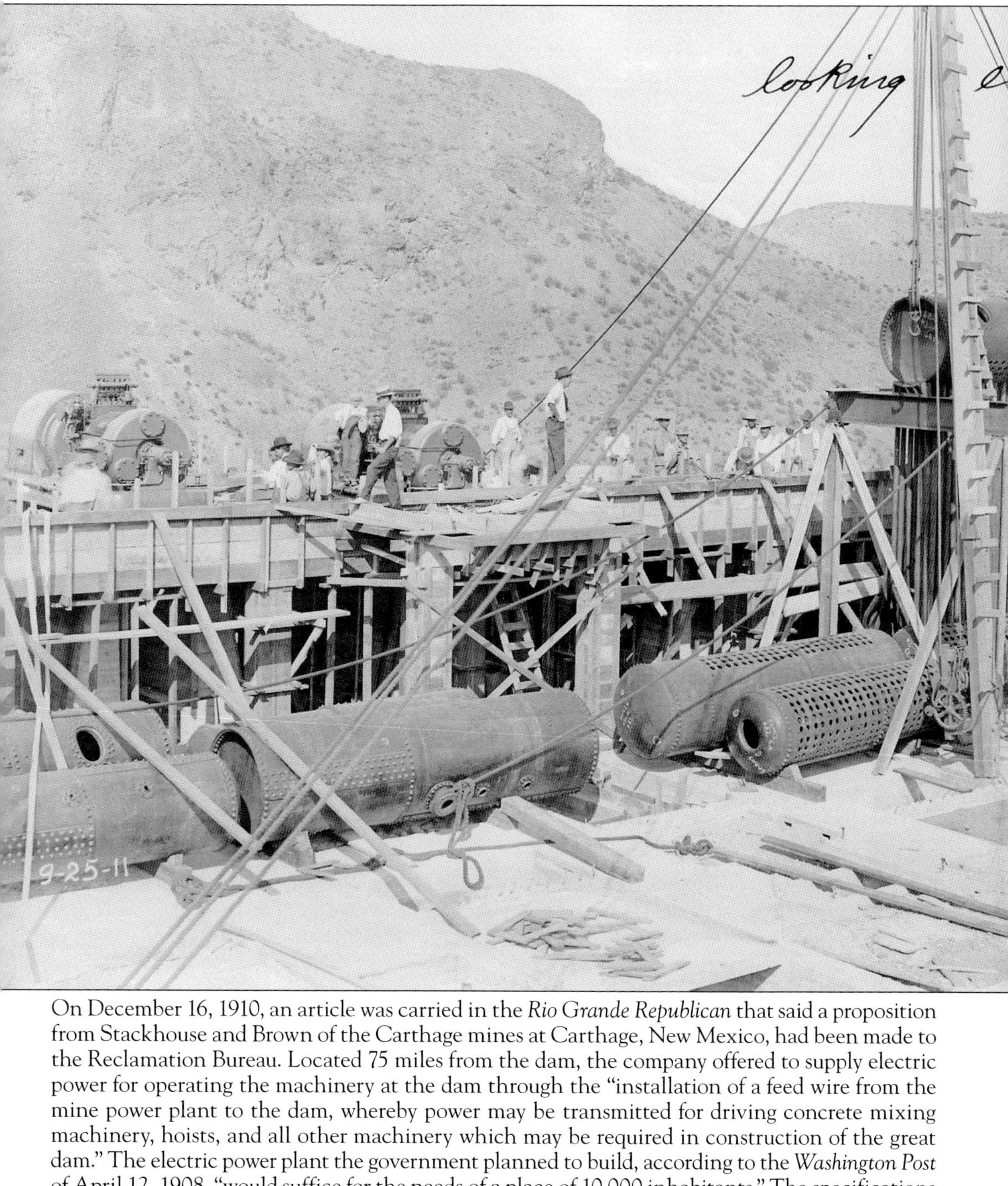

On December 16, 1910, an article was carried in the *Rio Grande Republican* that said a proposition from Stackhouse and Brown of the Carthage mines at Carthage, New Mexico, had been made to the Reclamation Bureau. Located 75 miles from the dam, the company offered to supply electric power for operating the machinery at the dam through the "installation of a feed wire from the mine power plant to the dam, whereby power may be transmitted for driving concrete mixing machinery, hoists, and all other machinery which may be required in construction of the great dam." The electric power plant the government planned to build, according to the *Washington Post* of April 12, 1908, "would suffice for the needs of a place of 10,000 inhabitants." The specifications for that power plant called for the installation of three steam turbine generators. The photograph above shows the first power plant under construction in 1911.

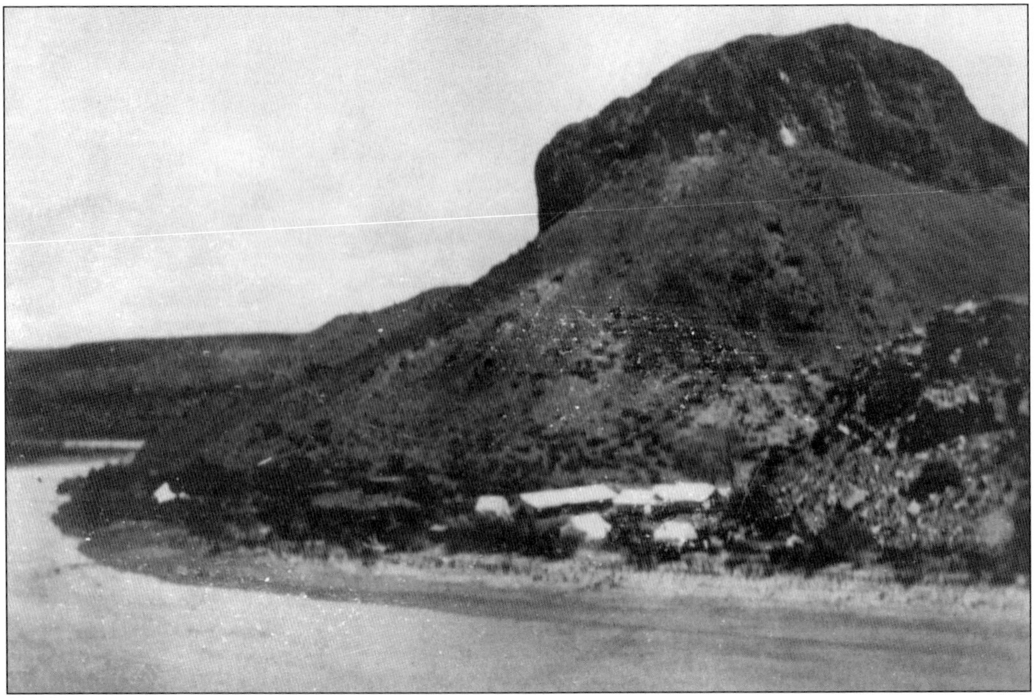

In 1911, a company expressed interest in developing 100 miles of electric line in the Rio Grande Valley. The current laws of 1911 only allowed for the lease of power sites on reclamation dams for 10 years, but Congress was looking into changing the law. This early photograph captures the landscape very early on in the construction of the dam.

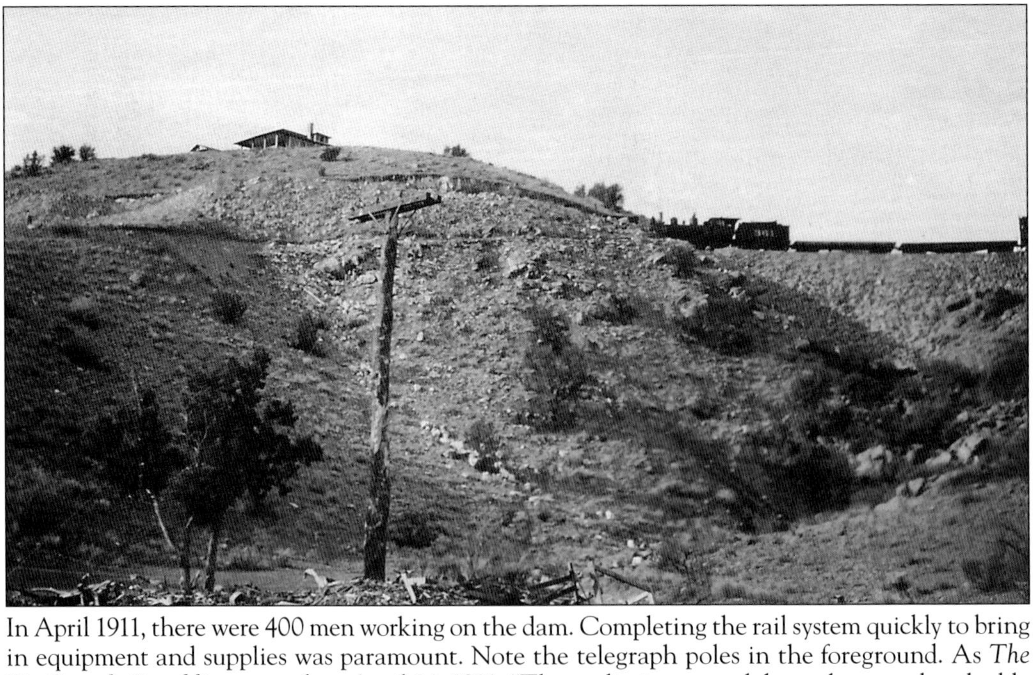

In April 1911, there were 400 men working on the dam. Completing the rail system quickly to bring in equipment and supplies was paramount. Note the telegraph poles in the foreground. As *The Rio Grande Republican* noted on April 14, 1911, "The preliminary work has taken much valuable time, but was very necessary in order to carry on the construction with economy and rapidity."

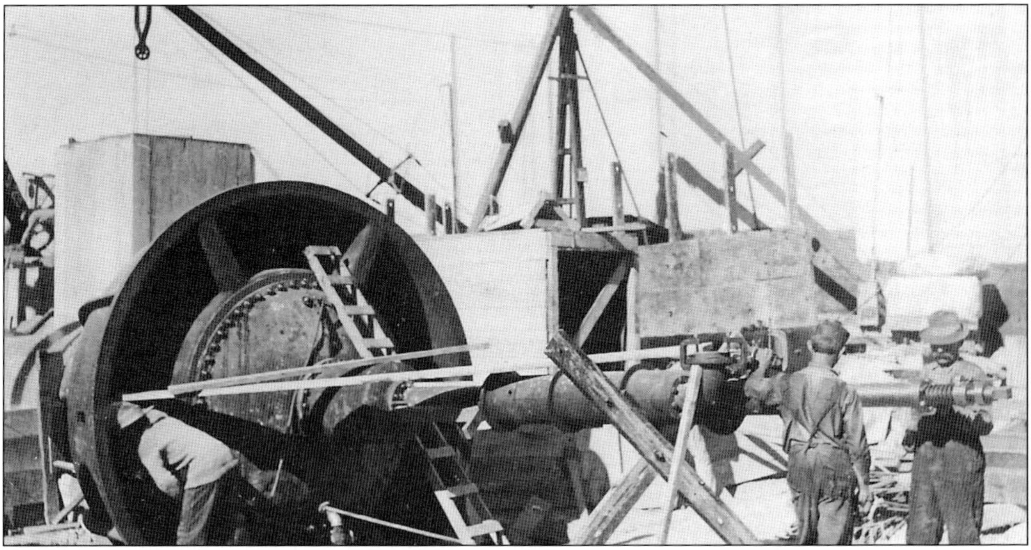

Throughout the construction, specialized equipment was being brought onto the damsite. Equipment was brought in packed inside wooden boxes by wagon and by rail. Pictured here are workers inspecting the recently arrived equipment; men from all over the country were arriving at the dam in anticipation of work.

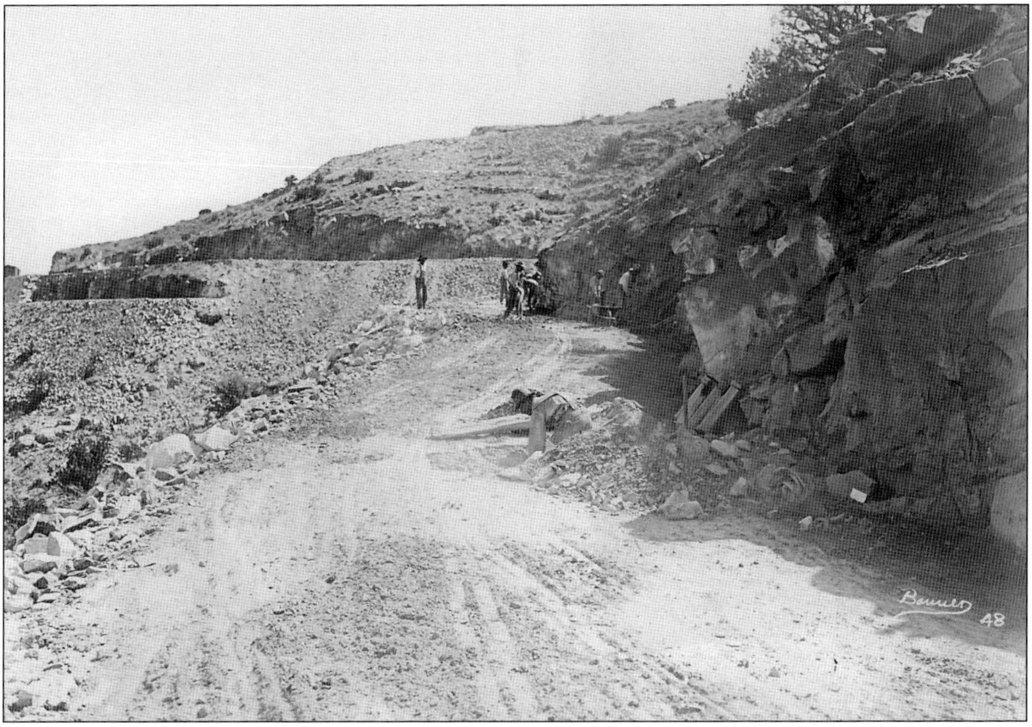

On April 14, 1911, the *Rio Grande Republican* reported that "Herbert W. Yeo, one of the junior engineers of the Elephant Butte Dam project, spent part of the week in Las Cruces, New Mexico, in the interests of the government." In discussing the project, Yeo said, "There is a force of some 100 men actually working on the Elephant Butte project." Many more men would be employed as the project continued.

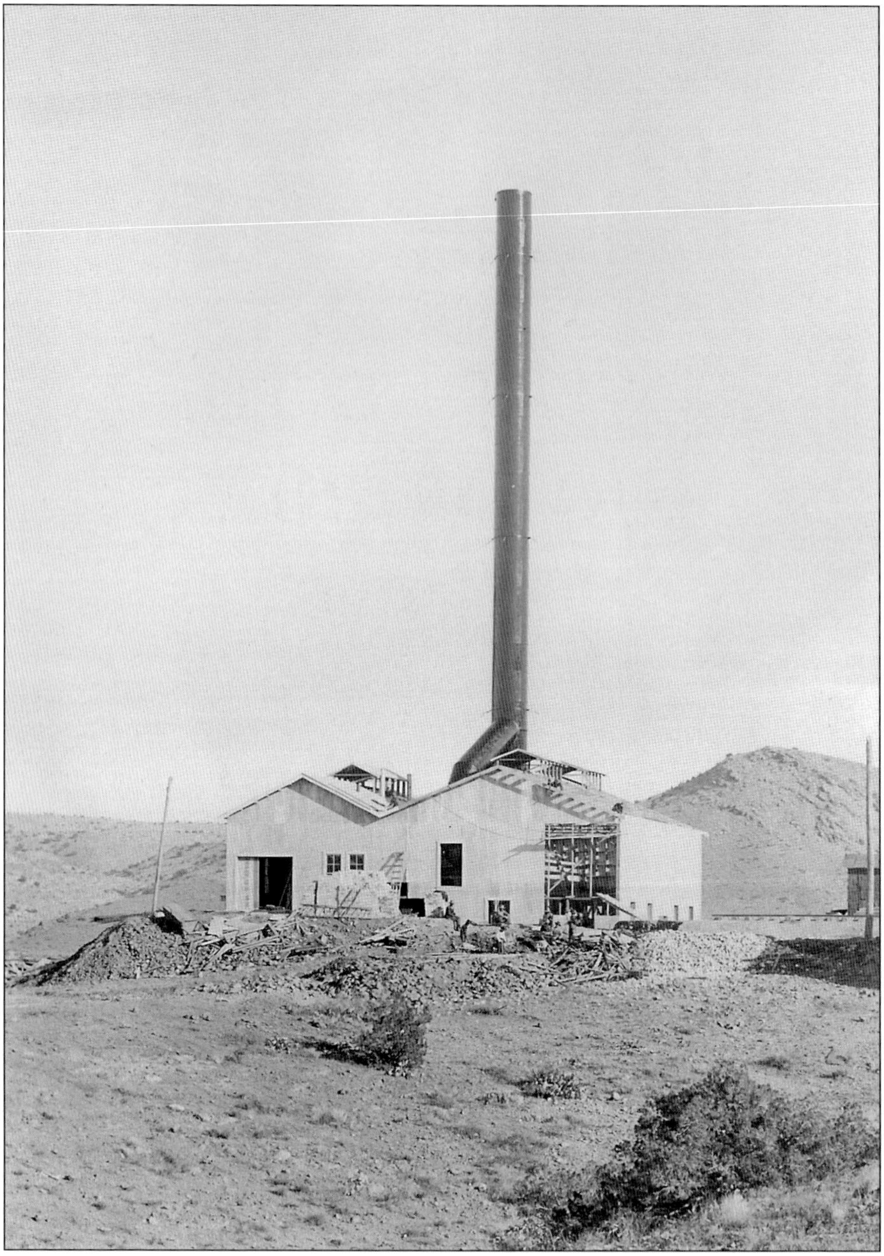

Pictured here is the smokestack and the sand-cement plant, in which finely ground sandrock was mixed with raw Portland cement. The sandrock was brought in from the quarry and passed through a small crusher, reduced from the size of a human head to the size of a large marble. If the sandrock was wet, it passed through, constantly revolving in a large cone-shaped dryer. At the large end of the dryer was a fire box, and at the small end was the smokestack. The fire and the heat from the fire box passed through a long revolving drum, tumbling the rock dry, and the shape of the drum caused the rock to work its way down to the large end, where it slowly dropped out onto an elevator. The heat that passed through the moving mass of the rock carried off the moisture. If the sandrock that passed through the crusher was dry, it passed by the dryer. This is just one of the phases of the concrete created for the Elephant Butte Dam construction project.

Two

Diverting the Rio Grande

The Flume

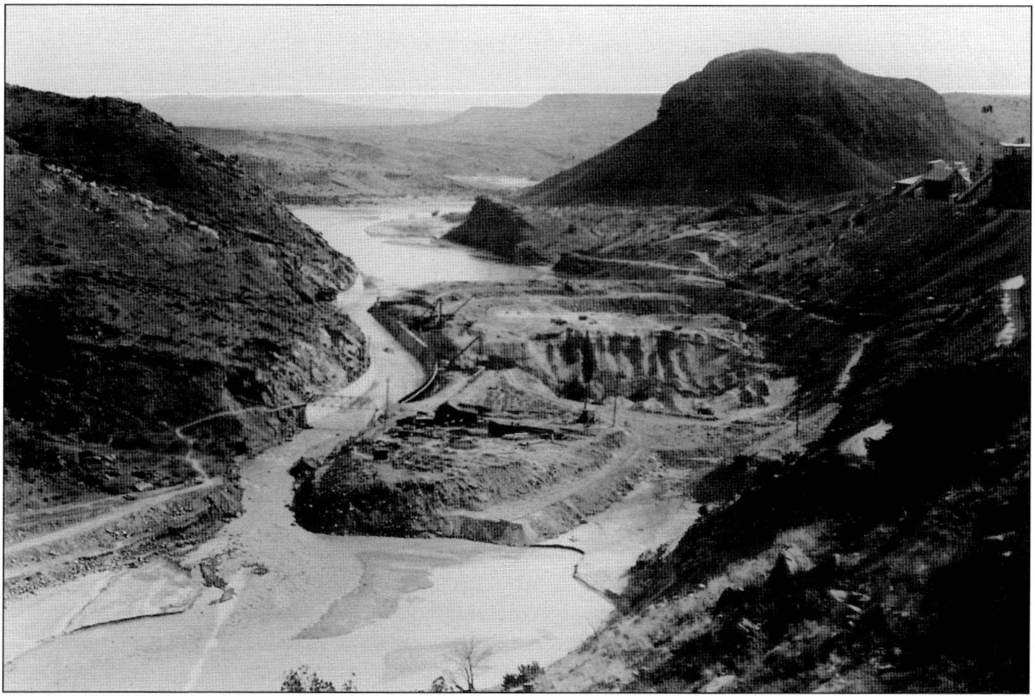

The flume, or diversion channel, had to be completed before construction of the dam could begin. The following photographs were preserved by Laverne Charles, acting construction engineer at the dam, and are referenced throughout this book. Enjoy the walk back into history.

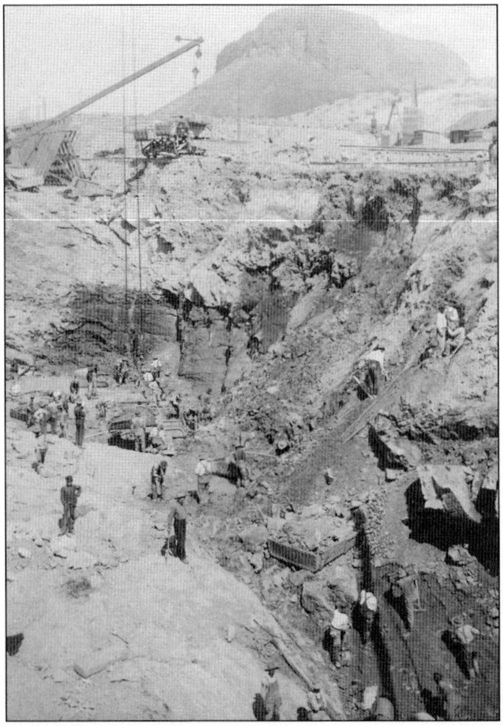

Pictured here is the excavation of the flume, which began in February 1911. The flume structure would actually be two parts, one portion being outside of the main dam structure and the other part being a portion of the dam. The material excavated for the flume was hauled off by narrow-gauge cars that were drawn by mules.

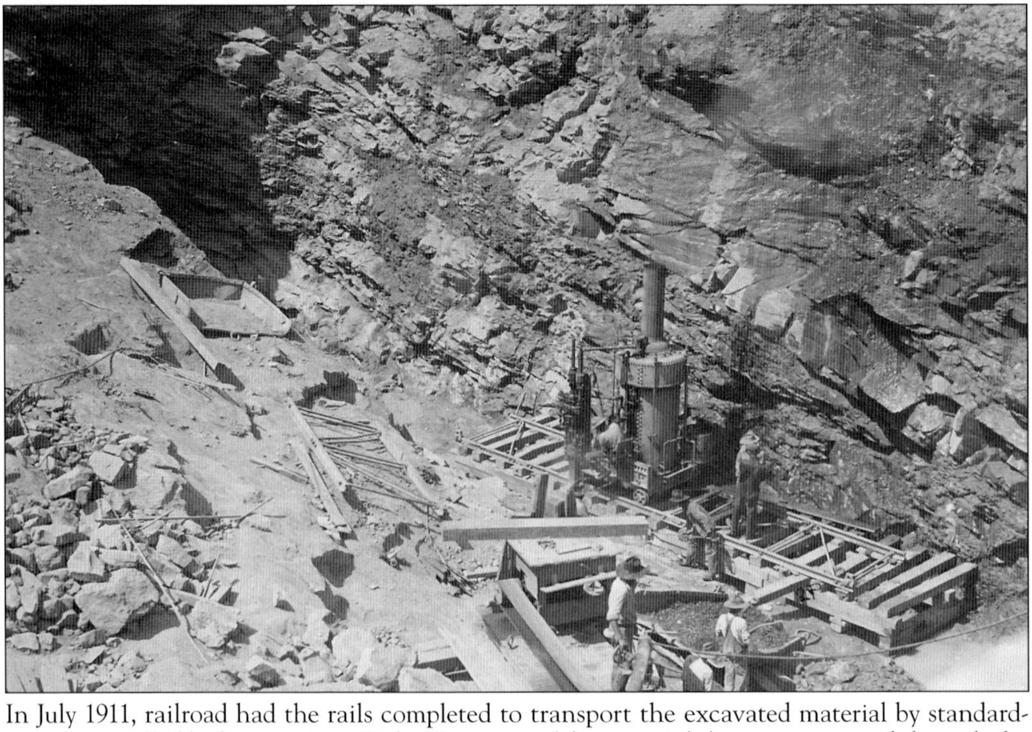

In July 1911, railroad had the rails completed to transport the excavated material by standard-gauge cars pulled by locomotives. Only 45 percent of the material that was recovered through the excavation was used in the building of the cofferdams, and 55 percent was considered to be waste material. They excavated to a maximum depth of 61 feet with drilling done by hand.

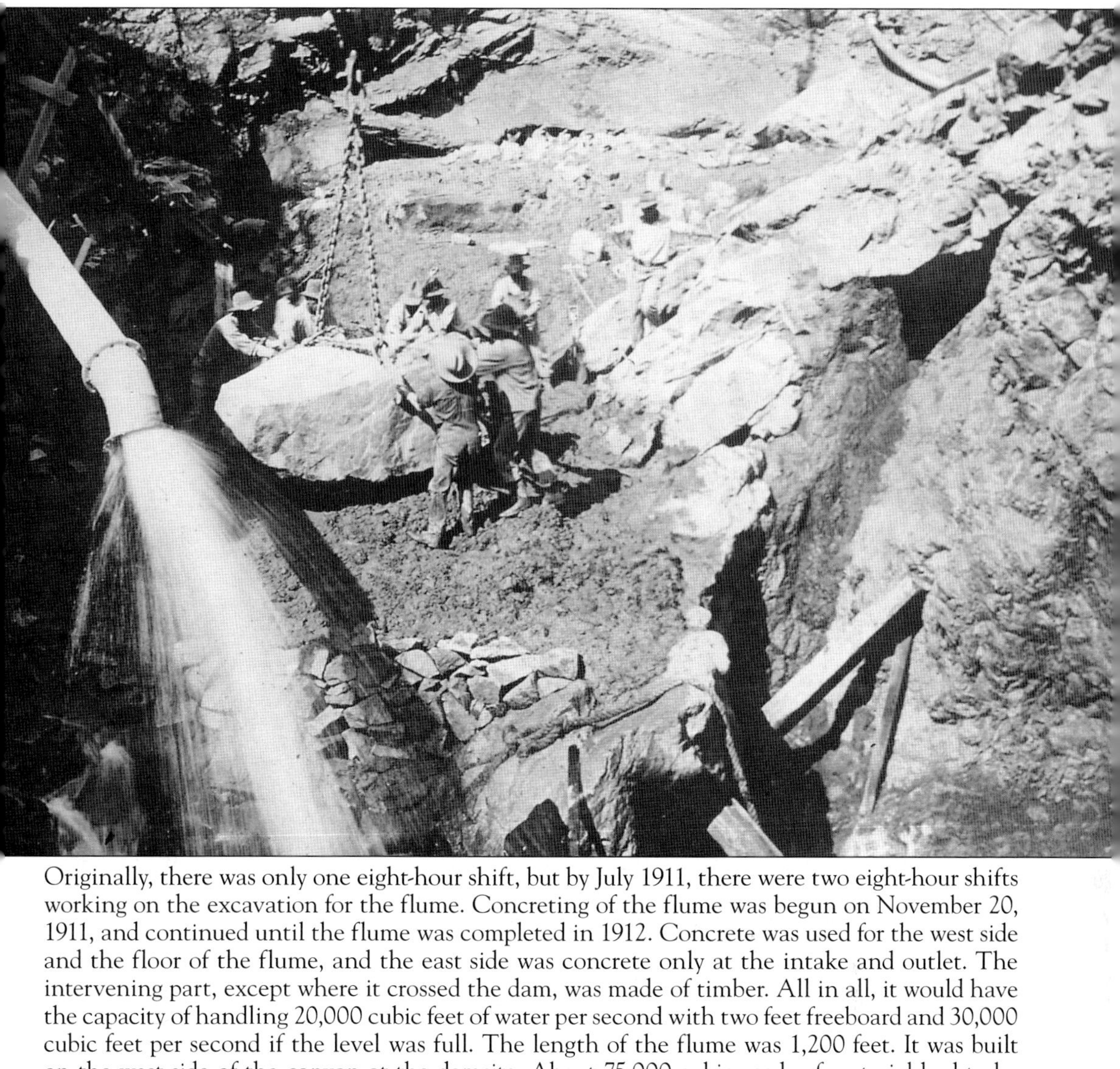

Originally, there was only one eight-hour shift, but by July 1911, there were two eight-hour shifts working on the excavation for the flume. Concreting of the flume was begun on November 20, 1911, and continued until the flume was completed in 1912. Concrete was used for the west side and the floor of the flume, and the east side was concrete only at the intake and outlet. The intervening part, except where it crossed the dam, was made of timber. All in all, it would have the capacity of handling 20,000 cubic feet of water per second with two feet freeboard and 30,000 cubic feet per second if the level was full. The length of the flume was 1,200 feet. It was built on the west side of the canyon at the damsite. About 75,000 cubic yards of material had to be excavated for the flume, and about 50,000 of that was solid rock.

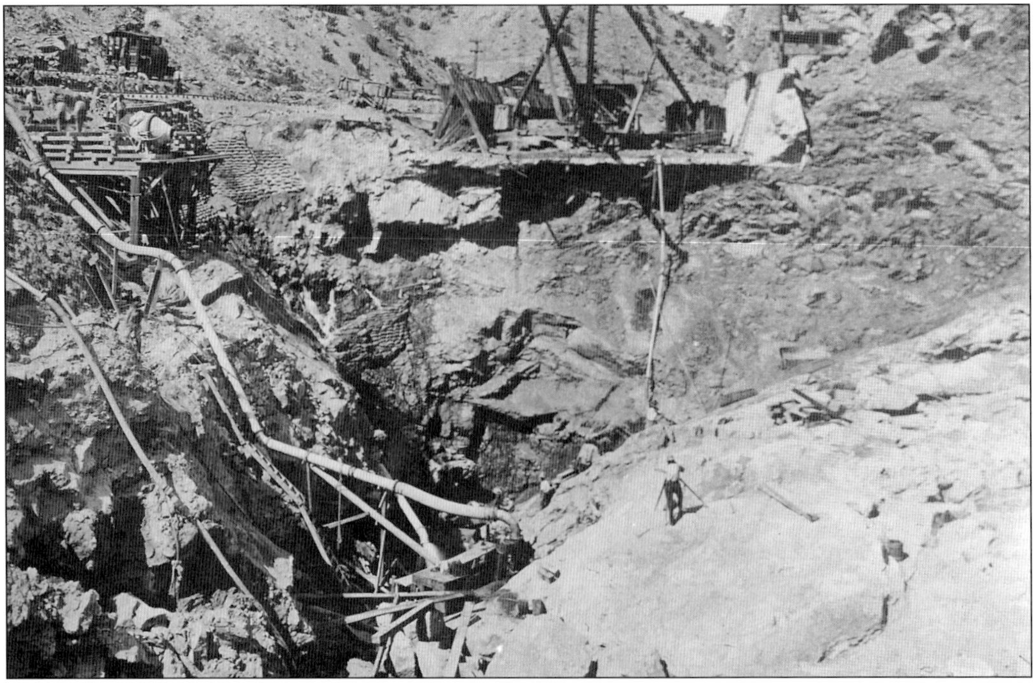

This photograph is marked as the excavation of the toe of the flume block in May 1912. State-of-the-art excavation equipment had been brought in along with both skilled and unskilled workers. The work was extremely dangerous, with steep slopes and loose from the deep cutting; one of the first buildings constructed on the damsite was the hospital.

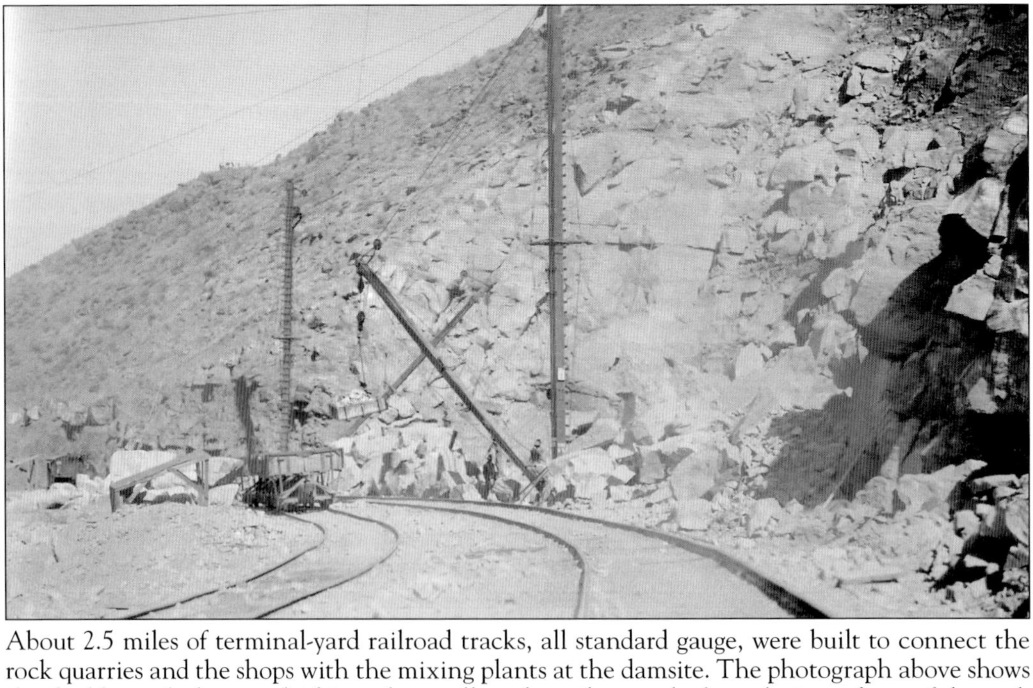

About 2.5 miles of terminal-yard railroad tracks, all standard gauge, were built to connect the rock quarries and the shops with the mixing plants at the damsite. The photograph above shows the double track that was laid in order to allow the railcars to be brought in and out of the rock quarries continuously.

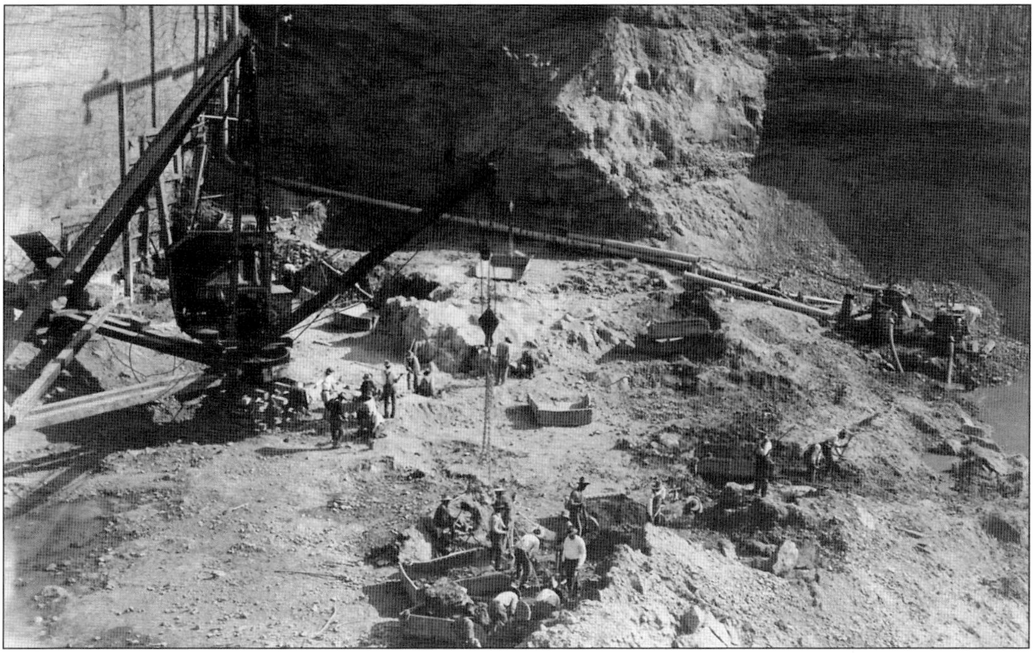

Pictured here in June 1912 are the derrick pedestals after a rockslide. At the time this photograph was taken, a bridge had been built across the river, a pump house was erected, telephone lines had been installed, permanent office buildings had been built, and dwellings for the workers had been completed.

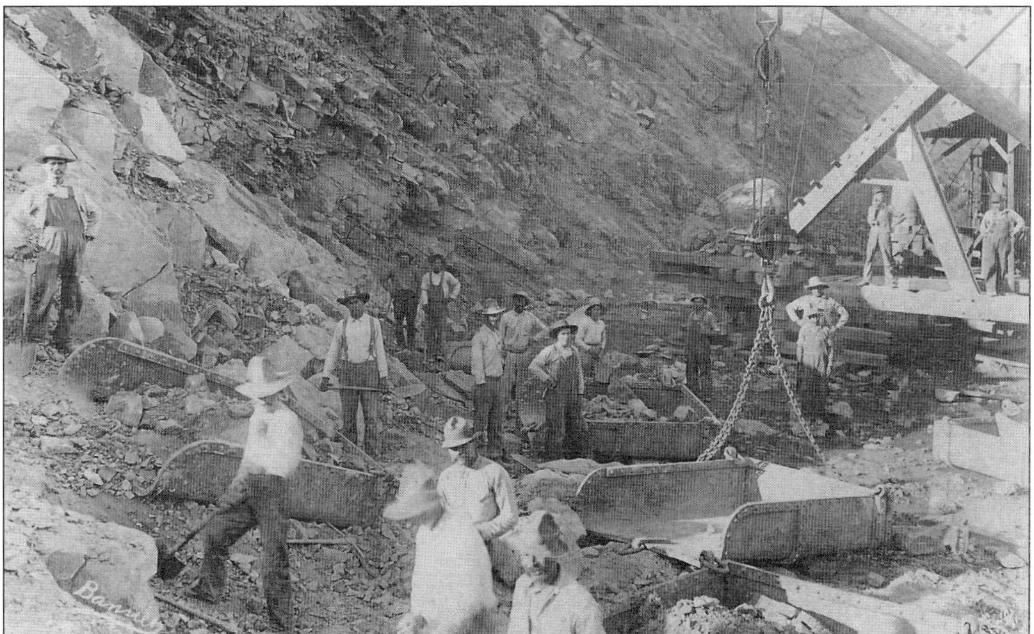

Workers pause for a photograph at the site of the flume blocks in July 1912. The concrete flumes were under construction, designed to carry the waters of the river past the damsite during the excavation and construction of the dam's foundations. By the end of 1912, the excavation of the flumes would be completed.

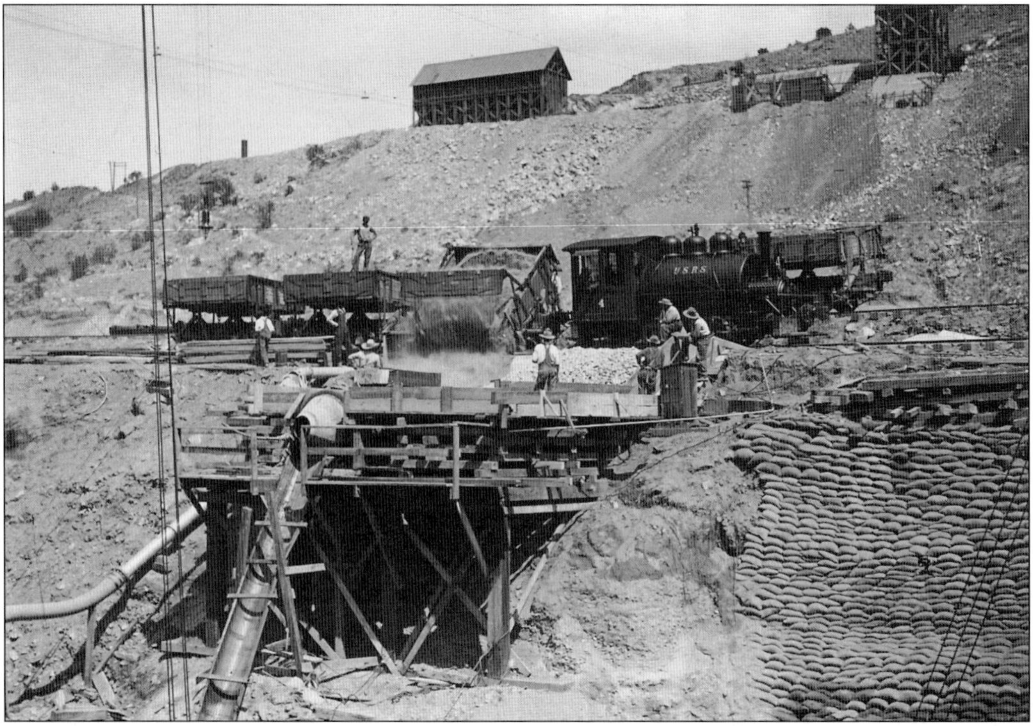

The Milwaukee Mixer, shown above, was used to mix the concrete in the construction of the flume. Milwaukee Mixers, one of the largest concrete mixers available at the time, boasted, "We will put a Milwaukee on your work. Test it out in your own way. Abuse it. Don't oil it. Don't block it up. And after using it for five days accept it or turn it down. You be the judge as to whether or not the Milwaukee is superior to any Mixer that you have ever used. All we ask is a chance to show you." Below are the Hotel Engle and the general merchandise store in Engle, New Mexico. Confusion over whether the town was spelled Engel or Engle was evident in the various spellings in the early newspapers.

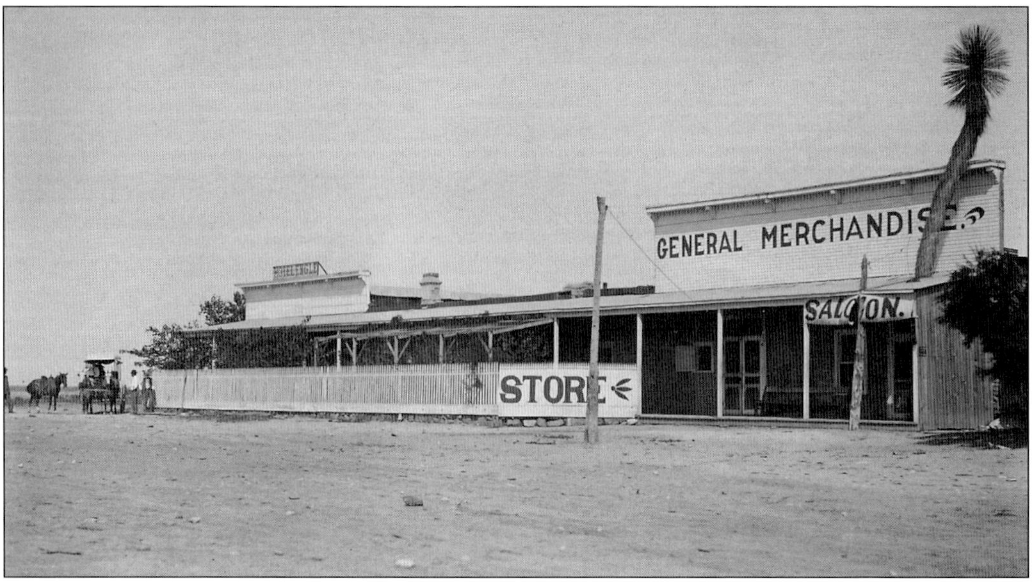

Junior engineer on the dam Herbert W. Yeo worked for the US Reclamation Service between 1908 and 1917, involved in the surveying and construction of the Elephant Butte Dam. The photograph here shows construction under Yeo's watch. Yeo would make his mark on New Mexico through his early photographs and documentation regarding the development of the highway system of New Mexico.

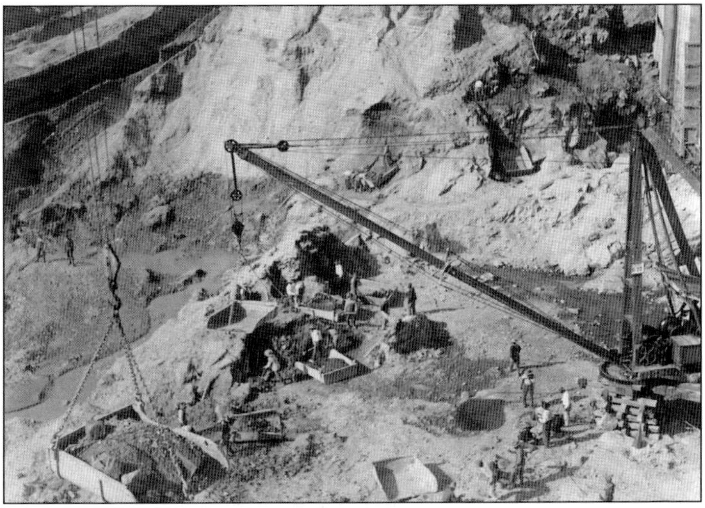

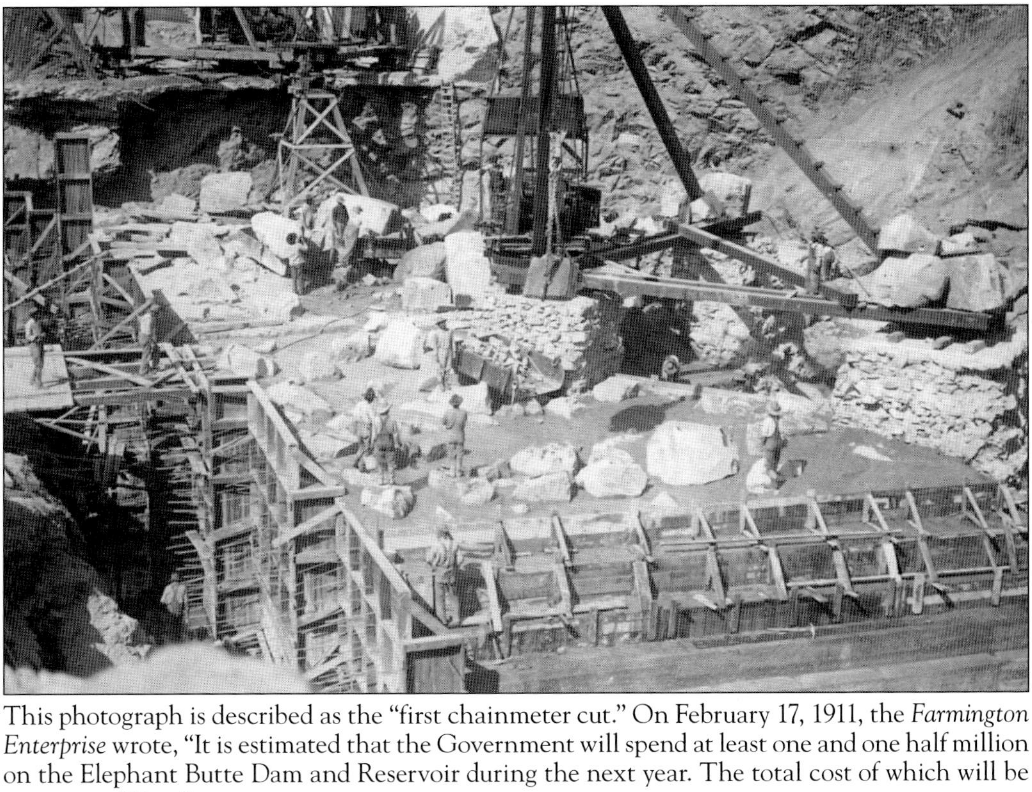

This photograph is described as the "first chainmeter cut." On February 17, 1911, the *Farmington Enterprise* wrote, "It is estimated that the Government will spend at least one and one half million on the Elephant Butte Dam and Reservoir during the next year. The total cost of which will be over ten million."

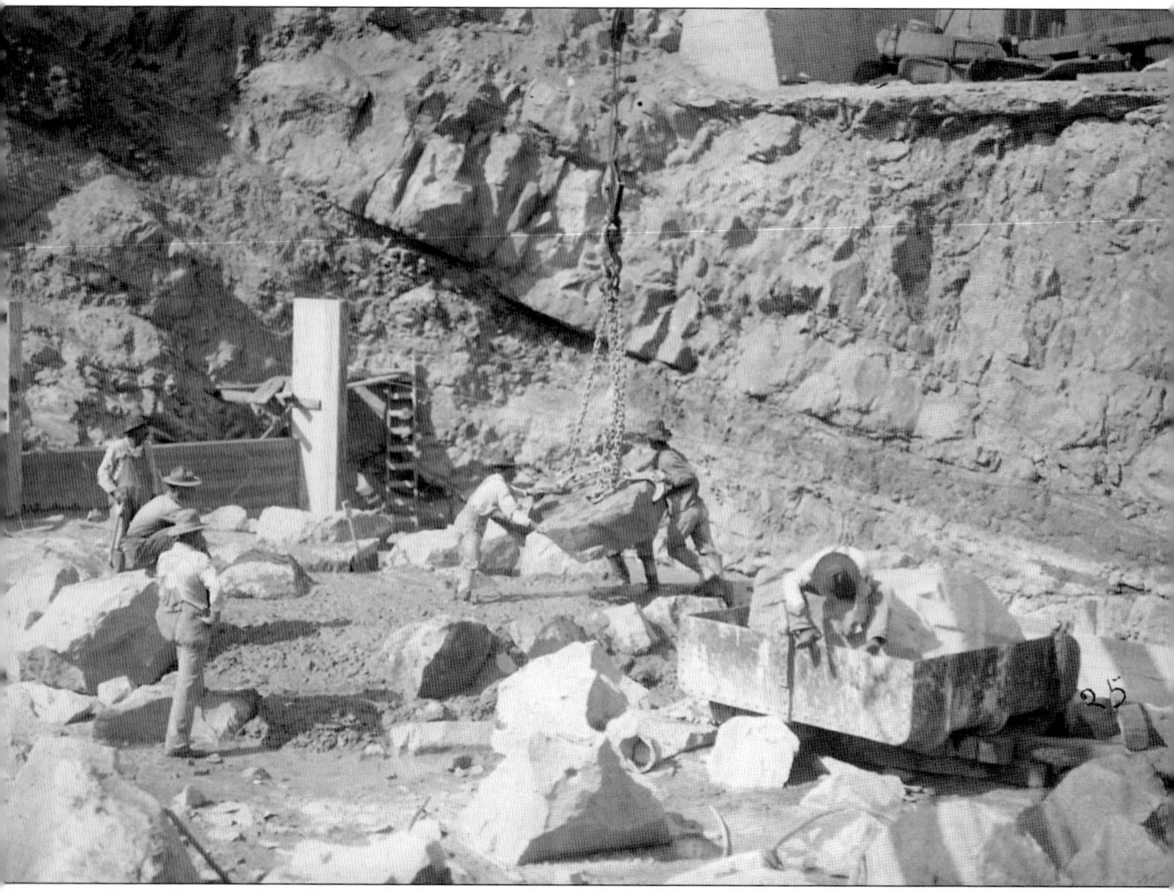

According to the *Albuquerque Morning Journal* from September 3, 1911, "Preparations are being made to waste a part of the materials excavated from the foundation of the flume and dump it into the river. This will form a part of the two cofferdams, which will be built to keep the water from the damsite during the time the foundations of the dam is being laid. One of these cofferdams will be 345 feet above the axis of the masonry dam and the other will be 188 feet below the axis of the afore-said dam. The upper one will be used to divert the water into the flume and the lower one will prevent the back water from flowing up the river on its exit from the flume. Men worked tirelessly during the excavation." The photograph is labeled, "Concreting in flume block looking up stream."

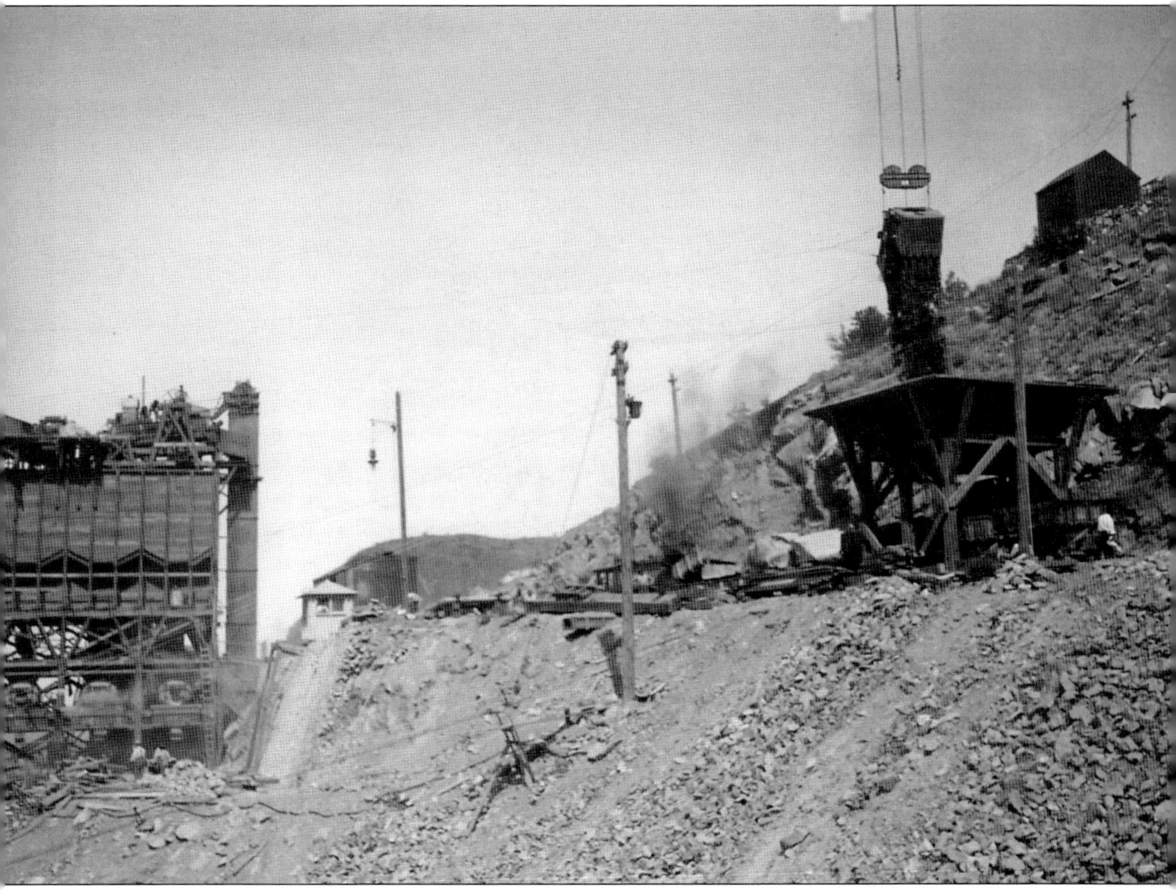

The sand-cement mixing plant had four floors. The fourth floor contained the bins for the sand, sand-cement, and crushed rock, and the third floor, attached to the ceiling, held the hoppers that measured the three ingredients for the concrete and worked with two gates. The top gate opened, the material filled the hopper from the bottom gate up, and then the top gate closed. Between the two gates was the desired quantity of measured sand, rock, or cement. The bottom gates opened, and the measured material fell into the mixers on the second floor. The water was then measured and put with the other materials. The gates were opened by water power controlled by a small valve. Mixers, which were large cast-iron hollow spheres that revolved on rollers, held nearly three yards of loose material per batch. The mixed concrete was dumped into a hopper over a narrow-gauge railroad track on the ground floor, and cars that were driven by cables hauled the mixed concrete out from under the main cableways in three-yard iron boxes, or skips.

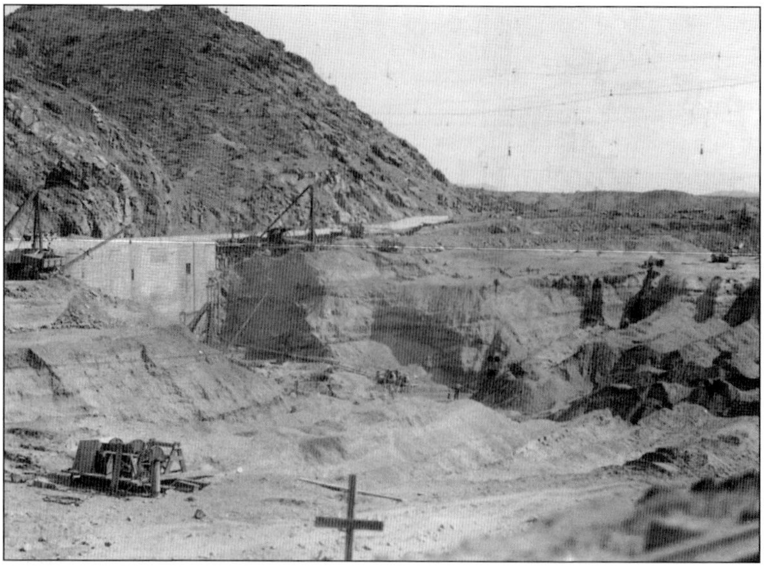

The *Rio Grande Republican* hailed the building of the Elephant Butte Dam on September 24, 1912, saying, "This great dam which holds back the waters of the Rio Grande during the flood time, and gives them out in time of drought, is of gigantic proportions and magnificent demonstration of engineering skill."

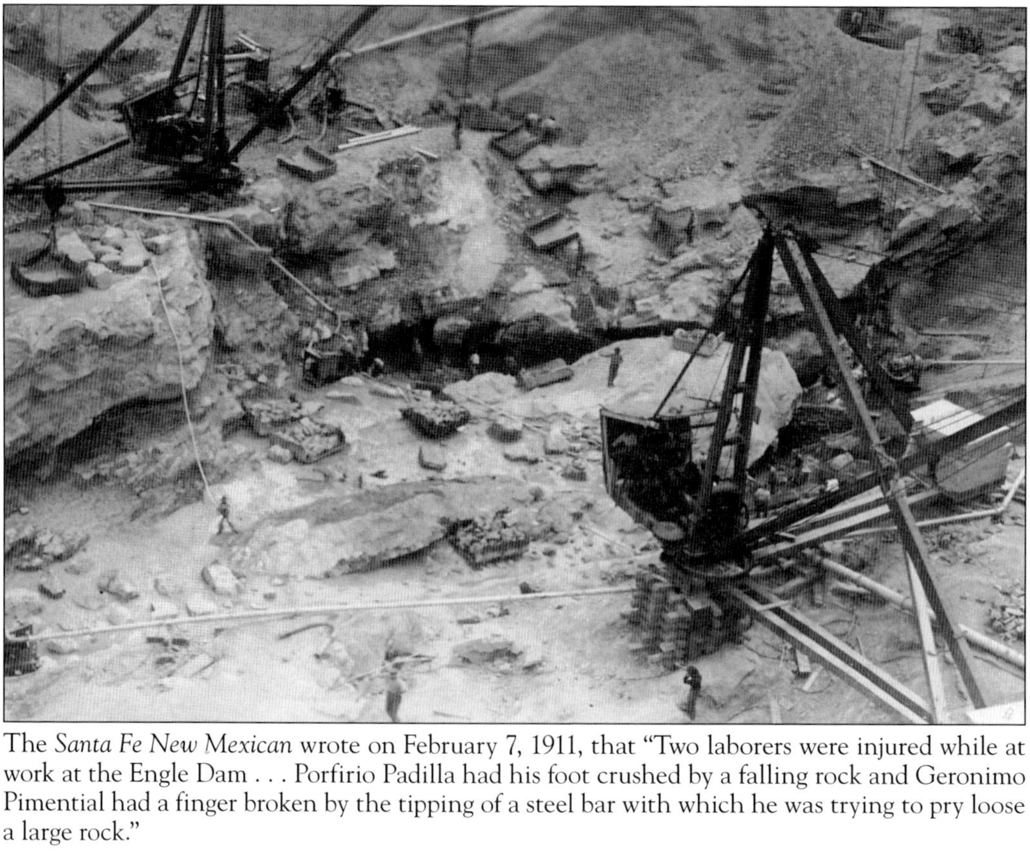

The *Santa Fe New Mexican* wrote on February 7, 1911, that "Two laborers were injured while at work at the Engle Dam . . . Porfirio Padilla had his foot crushed by a falling rock and Geronimo Pimential had a finger broken by the tipping of a steel bar with which he was trying to pry loose a large rock."

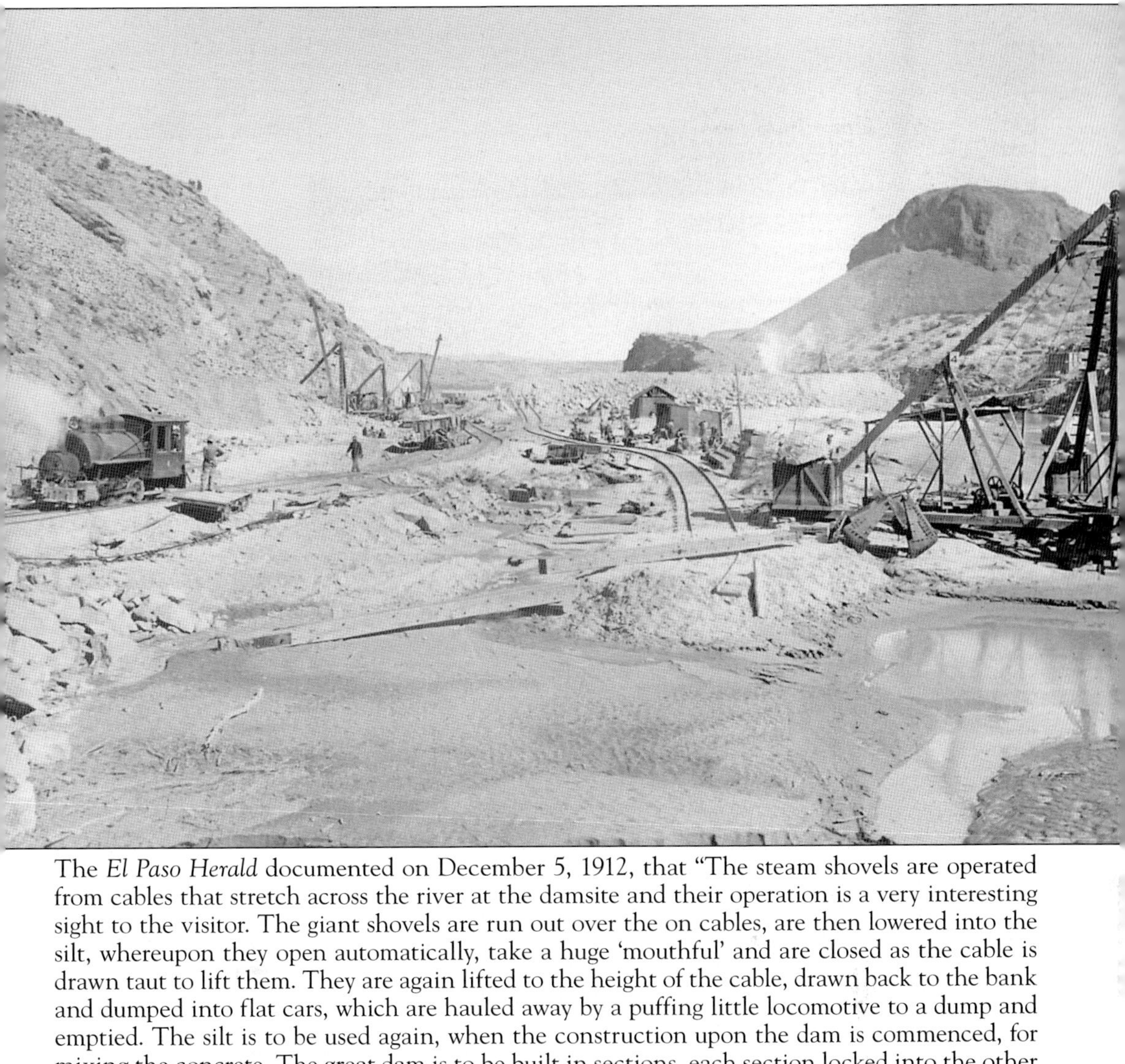

The *El Paso Herald* documented on December 5, 1912, that "The steam shovels are operated from cables that stretch across the river at the damsite and their operation is a very interesting sight to the visitor. The giant shovels are run out over the on cables, are then lowered into the silt, whereupon they open automatically, take a huge 'mouthful' and are closed as the cable is drawn taut to lift them. They are again lifted to the height of the cable, drawn back to the bank and dumped into flat cars, which are hauled away by a puffing little locomotive to a dump and emptied. The silt is to be used again, when the construction upon the dam is commenced, for mixing the concrete. The great dam is to be built in sections, each section locked into the other with giant concrete teeth and steel reinforcing, so that contraction and expanse are allowed for. Otherwise, if the dam were built in a solid piece, there might be a cracking."

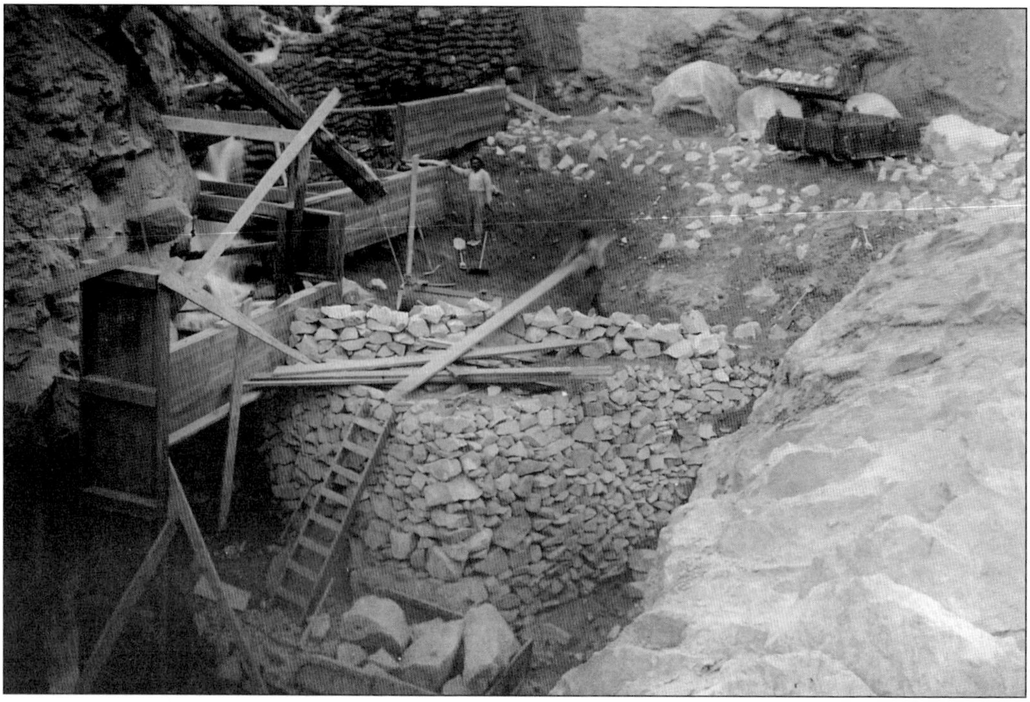

In the April 12, 1908, edition of the *Washington Post*, R.C. Willing of Santa Fe, New Mexico, was quoted as saying, "For the next six years Uncle Sam will be working day and night to complete what will be the biggest dam and store reservoir in the world." New Mexico at this time was still a territory and would not become a state until 1912. Interestingly, R.C. Willing referred to the dam as the Elephant Butte Dam at Engle. Most of the newspapers at that time were referring to the future dam as Engle Dam. It was estimated in 1908 that the future dam would irrigate 180,000 acres of land in Texas, New Mexico, and Mexico. The article went on to say that the "artificial lake" would be 40 miles long and could not supply water to a city the size of Paris for 10 years.

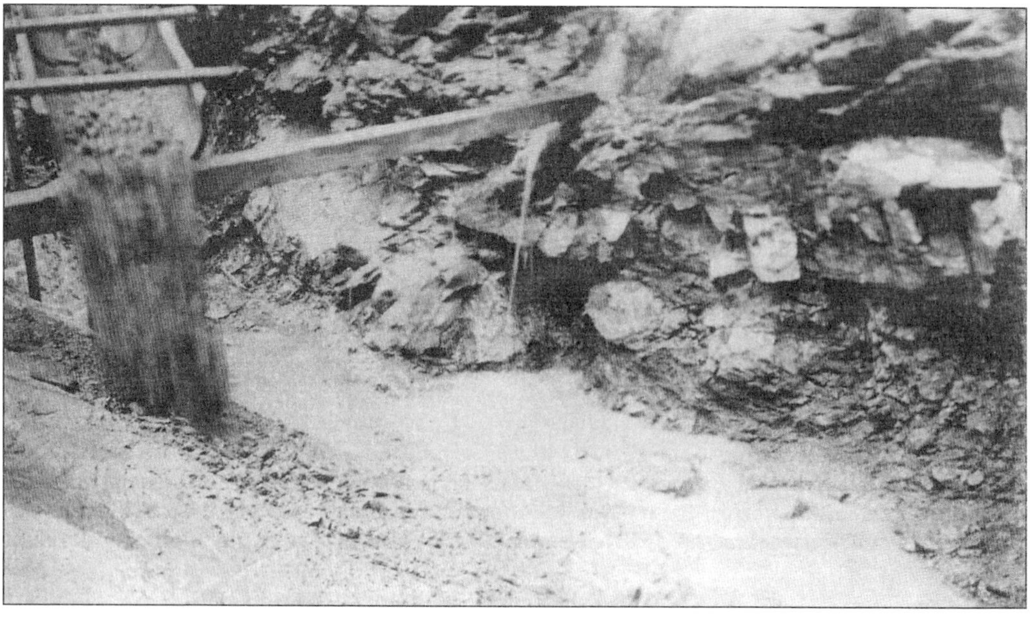

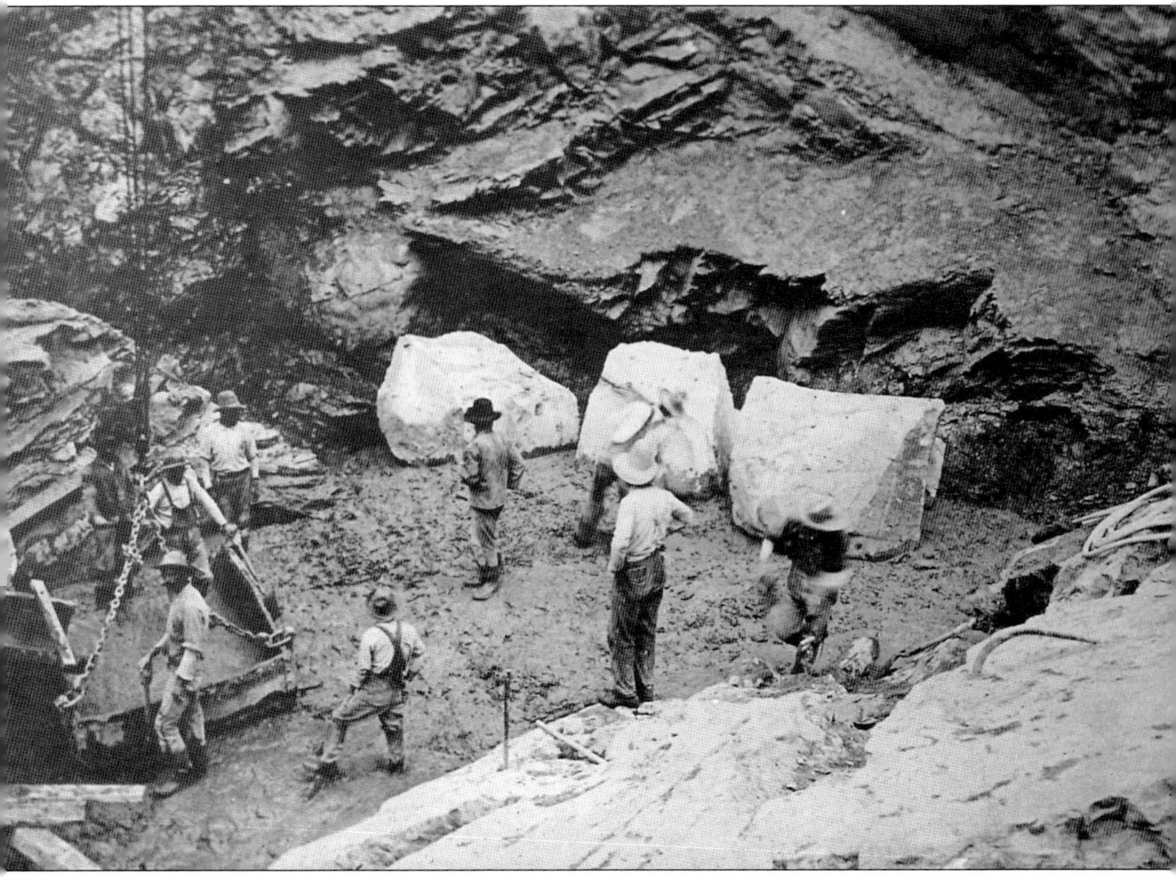

Albuquerque Journal headlines on November 21, 1911, declared the "Government Determined That Nothing Shall Be Allowed to Interfere With Progress of Work On Great Elephant Butte Reclamation Project." John Wyler was said to have done the boiler work, and W.J. Heinritz, a representative of the General Electric Company, had installed the switch board for the plant. An appropriation of $6,355,000 was given in 1911 to help with construction costs and the cost of the irrigation structures that would be downstream from the dam. Two diamond drills were excavating the foundation of the dam by January 11, 1911, and on average, two carloads of cement from El Paso were being delivered weekly. This photograph was taken in 1912, and with the exception of the locomotives, electrical power ran all parts of the damsite. It was said that the mechanical shops were so well equipped that they included a blacksmith, machine, and woodworking shop that could not only repair but build any part necessary or needed.

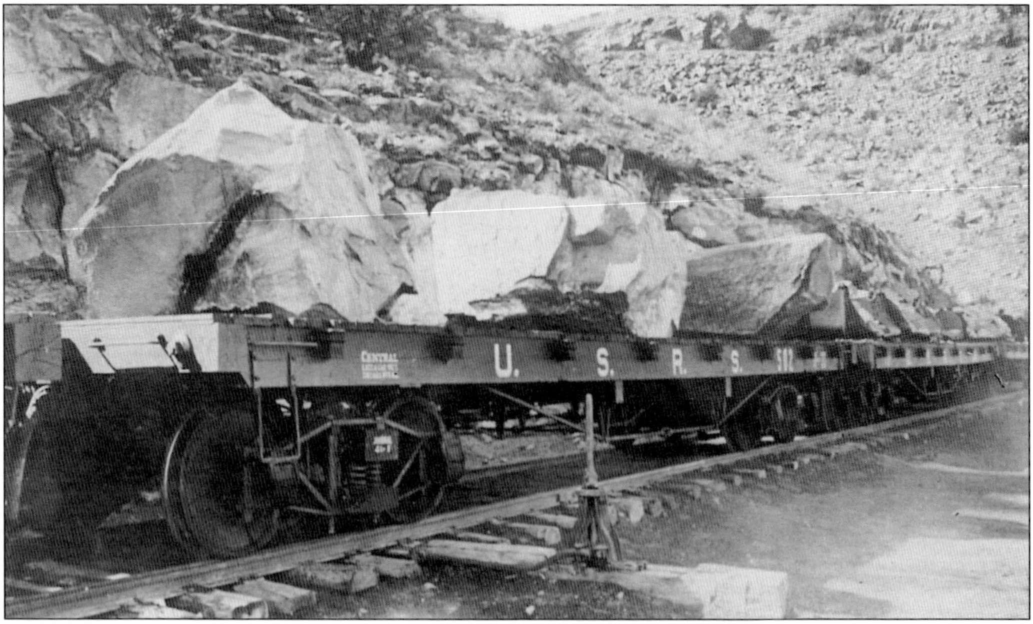

The *Rio Grande Republic* of January 29, 1915, reported Elephant Butte was "of the gravity type. It is building of cyclopean concrete, which being interpreted, means that it is composed of the usual concrete mixture poured around huge rocks weighing three or four tons each and conveyed from a quarry about a mile back in the hills to be dropped, like plums in a pudding, all through the gigantic dam."

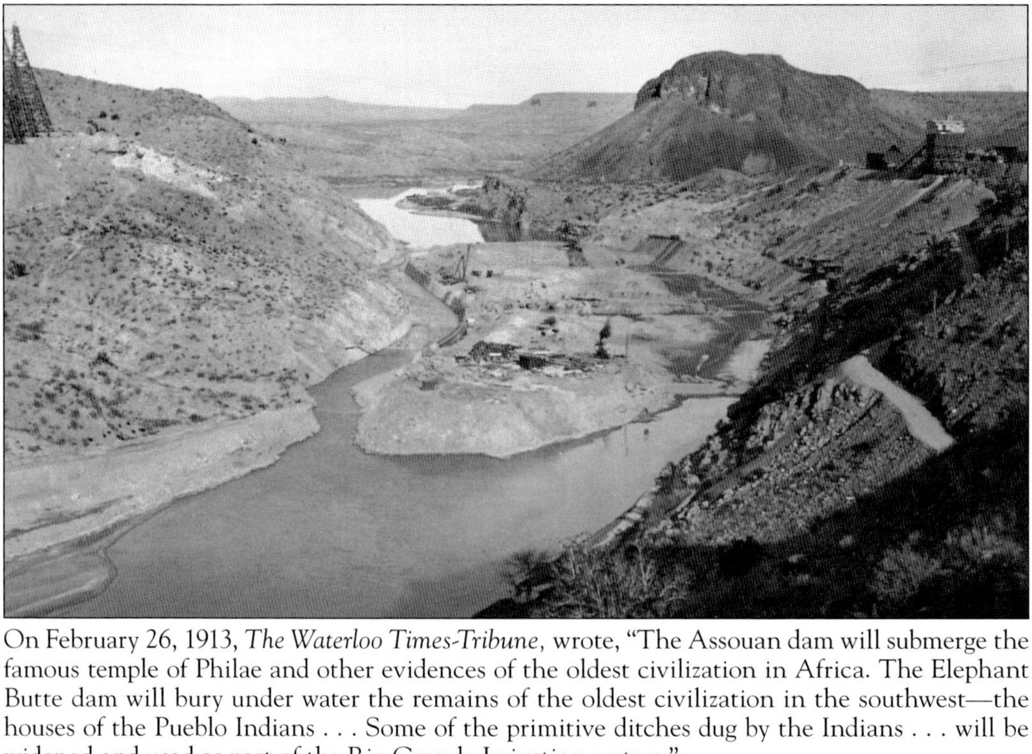

On February 26, 1913, *The Waterloo Times-Tribune,* wrote, "The Assouan dam will submerge the famous temple of Philae and other evidences of the oldest civilization in Africa. The Elephant Butte dam will bury under water the remains of the oldest civilization in the southwest—the houses of the Pueblo Indians . . . Some of the primitive ditches dug by the Indians . . . will be widened and used as part of the Rio Grande Irrigation system."

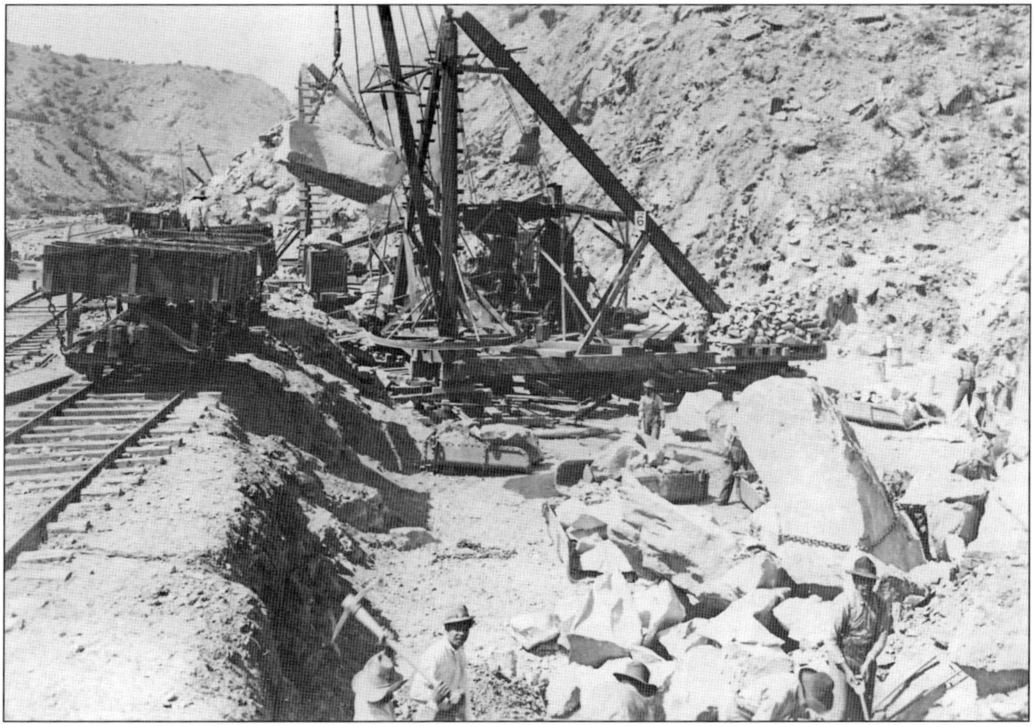

Plum stones (large boulders) were transported from the quarry by the railroad as seen in this photograph from June 2012. In all, plum stones accounted for about 15 percent of the total volume of the dam. They were imbedded by teams of men who also shoveled the concrete up against the forms, and the equipment capable on loading these plum stones had to be shipped and then assembled at the damsite. Rock quarries had been opened and were equipped with electric derricks to this end. The railroad had been built with double tracts from the quarries to the damsite, which was a distance of one mile. By 1912, most of the preliminary work for the construction had been completed, but in the image below, flume blocks are still being constructed.

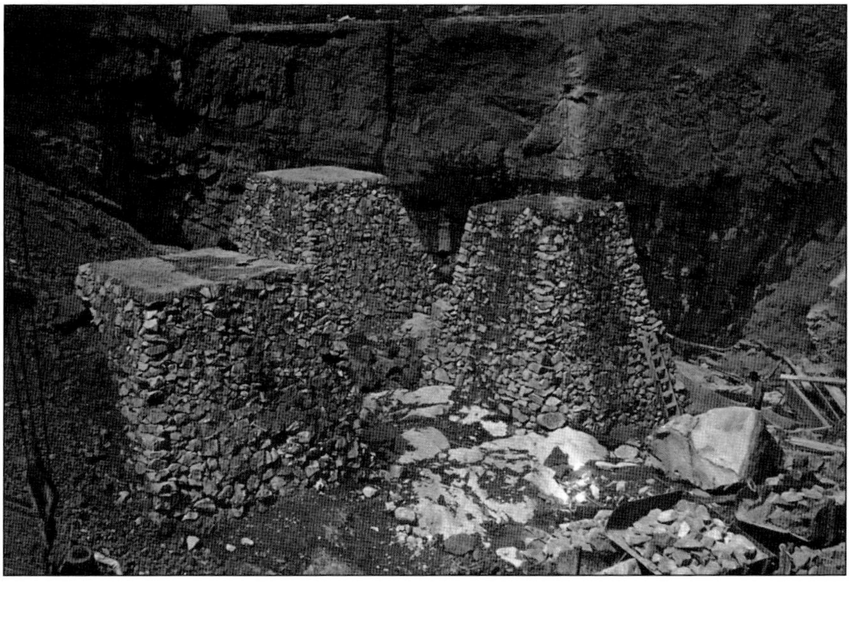

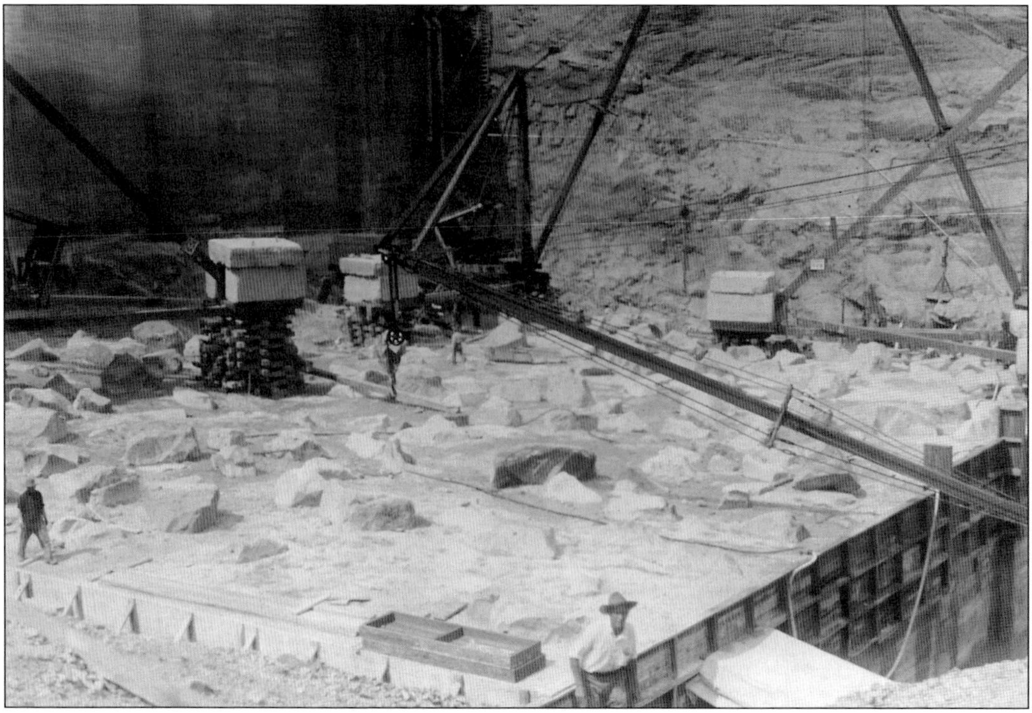

Three cableways had been erected over the damsite to convey materials and remove excavation from the flume section. The newspapers of the surrounding area reported on the construction of the dam, and interest was high in the actual construction of this gigantic undertaking by the federal government. On February 16, 1912, the *Santa Fe New Mexican* reported on Francis E. Lester's trip from Mesilla Park to Elephant Butte and back. The car was driven by his son, Ted Lester, and Joseph Taylor and W. Strode of Doña Ana, New Mexico, were guests. "This is the first automobile trip ever made between Las Cruces and Elephant Butte. Mr. Lester reports a splendid road for the greater part of the way."

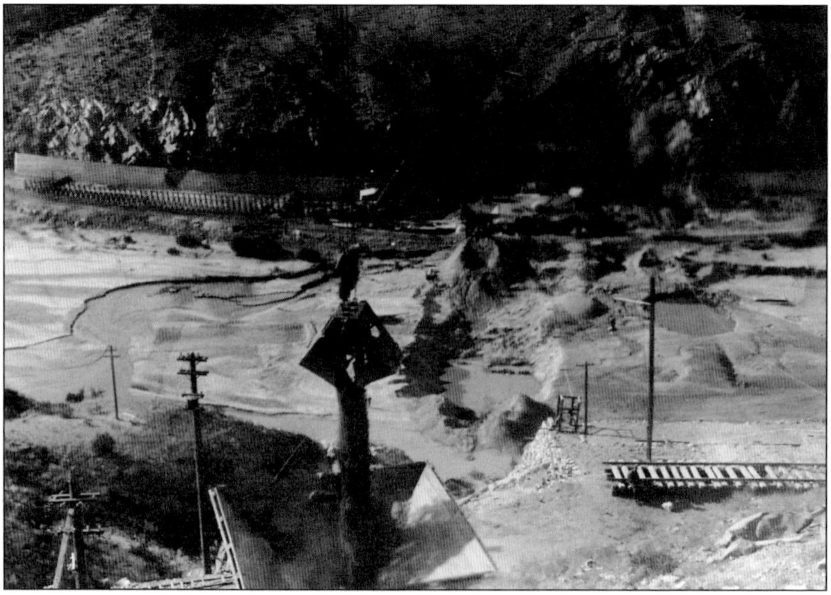

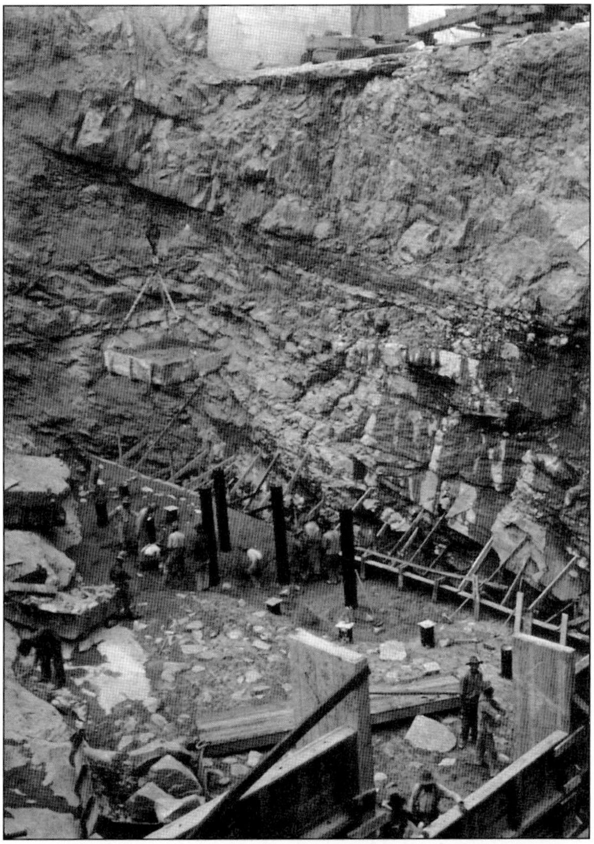

New Mexico was part of the Spanish Empire for over 200 years and part of the country of Mexico for another 25 years. In 1948, the United States obtained the Territory of New Mexico as part of the Treaty of Guadalupe Hidalgo. It took until January 6, 1912, before the US Congress would admit New Mexico as the 47th state in the Union. Construction on the Elephant Butte Dam project had begun in 1908 but was suspended due to issues with land acquisition. By 1912, construction had resumed. On May 10, 1912, the *Farmington Enterprise* linked statehood with the "opportunity to make New Mexico a great state, known to the world for its superior advantages as a place for investment, pleasure, and health" and promised that the lake created by the construction of the dam would "afford one of the most beautiful and attractive summer resorts and fishing grounds in America."

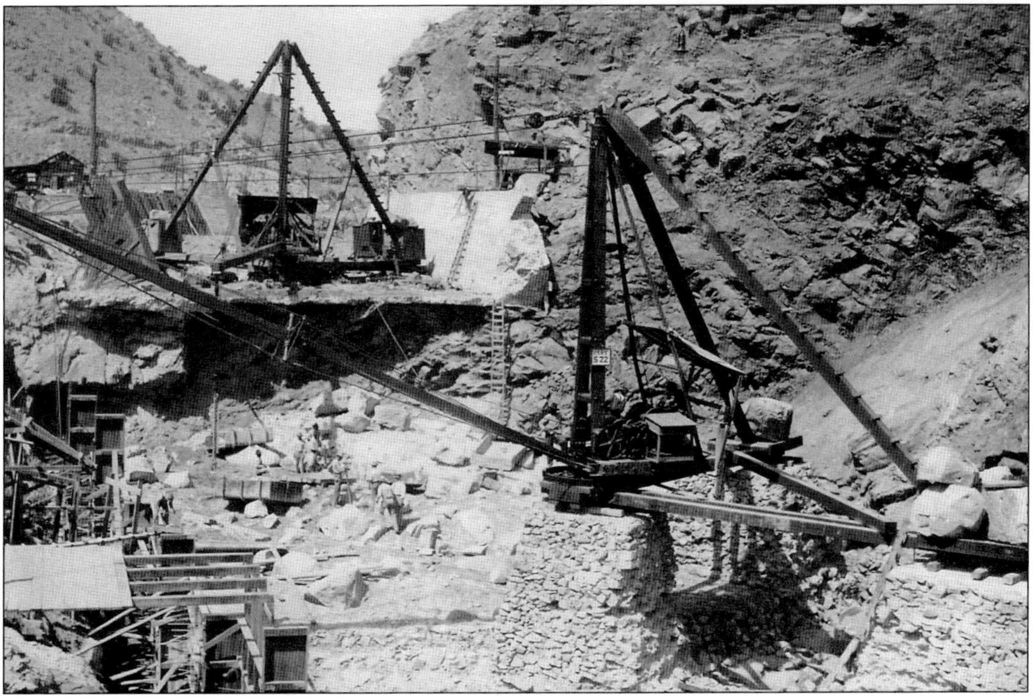

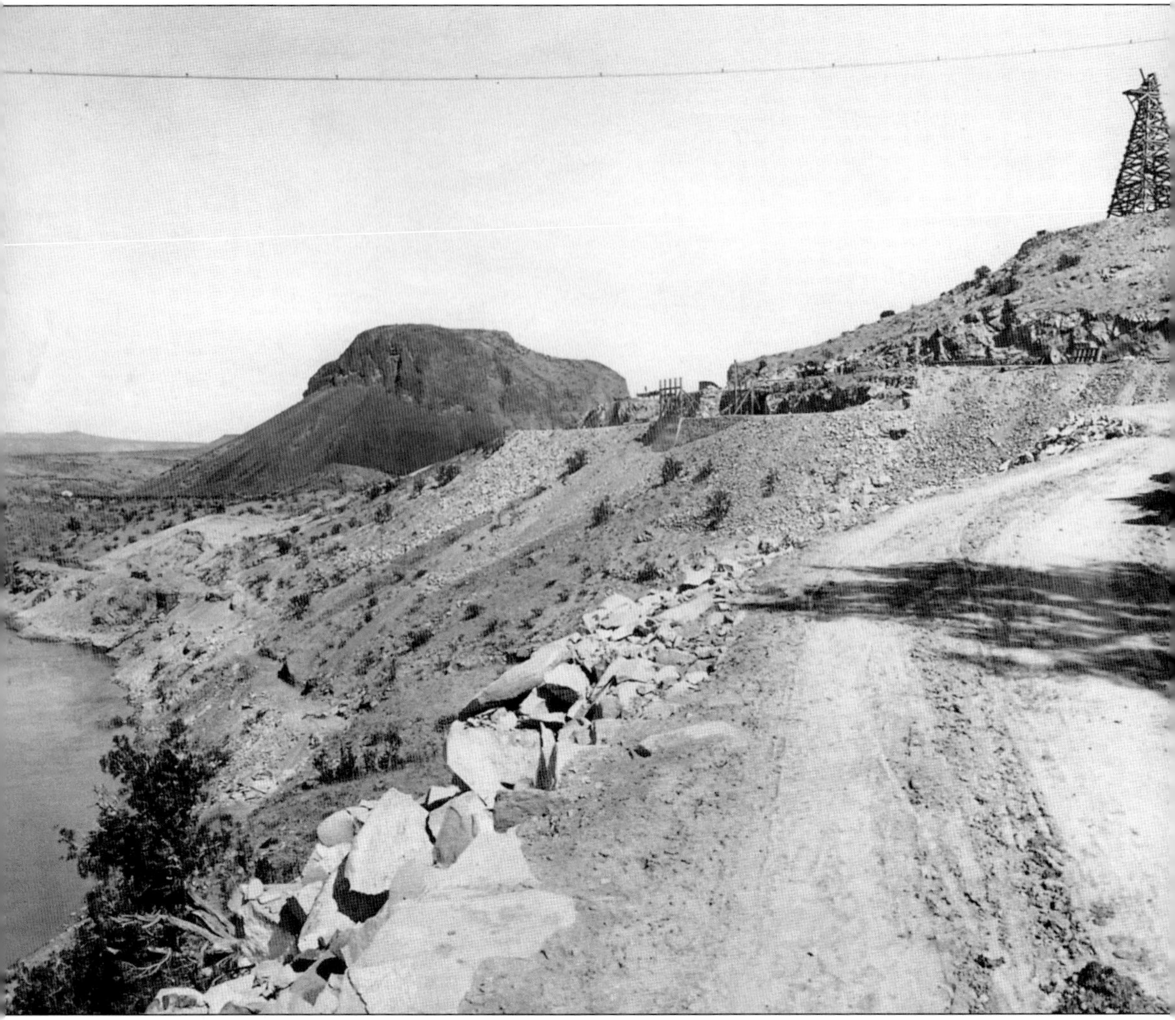

The photograph above shows the towers from which the three Lidergerwood cableways, used to handle heavy materials, were strung across the damsite. Each of these three cableways was capable of carrying eight tons, the cables were 2.25 inches in diameter of the locked-strand type, and they had about a 1,400-foot span. Two of the cables were made by the Trenton Iron Works, and one was made by A. Leschen & Sons Company. The cableways had a hoisting speed of 200 feet per minute and a traveling speed of 800 feet per minute. It was cautioned that it would be unwise for an operator to increase the speed on account of the relatively short haul. Each cableway was operated by a 300-horsepower motor. Cableways were not a new invention at this time, as they had been used in the construction of earlier reclamation dams.

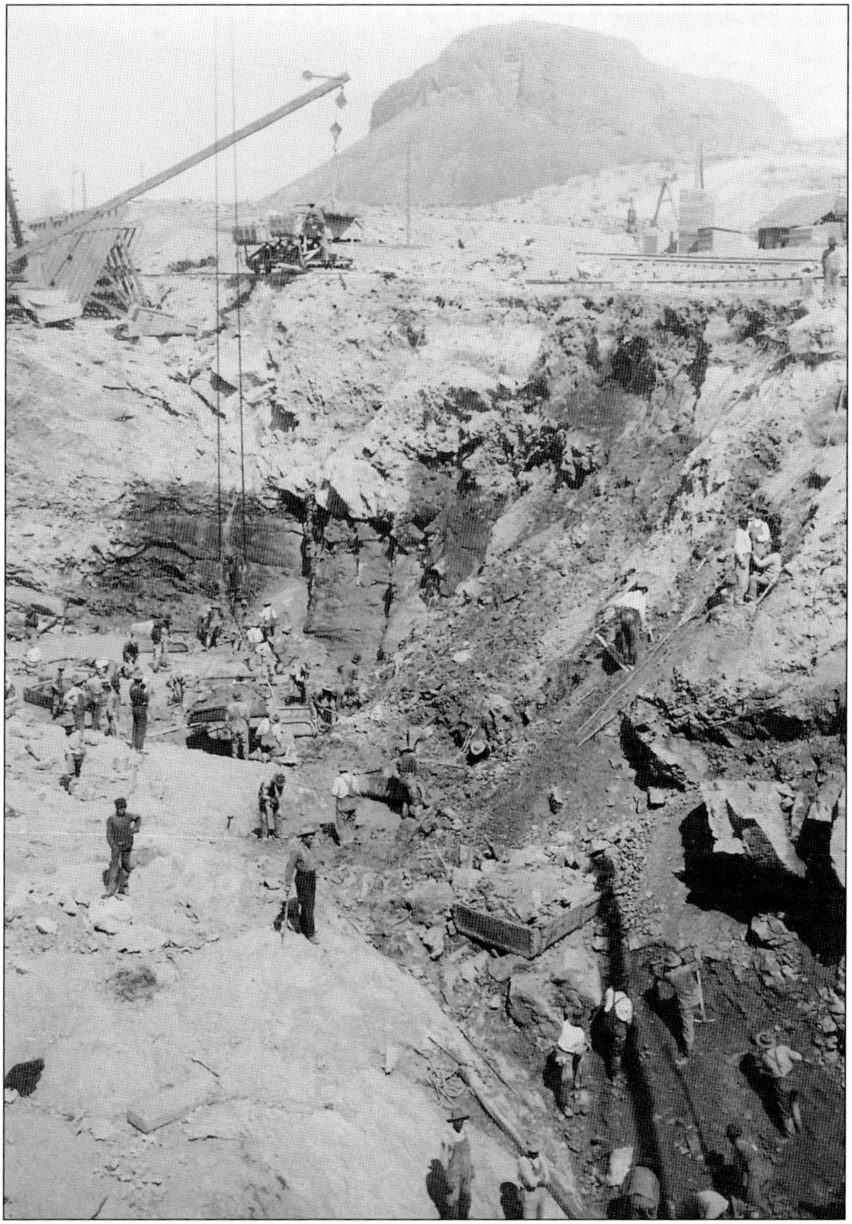

According to the *Washington Post* of July 6, 1913, E.H. Baldwin, consulting engineer, wrote, "During the latter part of April the grab buckets which had been excavating about 1,200 cubic yards of sand and gravel per day on the damsite, were removed from cableways and 'skipe' substituted in order to help out the derricks, which were handling boulders and loose rock on a section near the flume, the desire being to prepare an area of the foundation about 200 feet square, adjacent to the flume section for masonry, so that concreting could be going on while the balance of excavation was being completed, thus enabling a large force to be employed and consequently hasten the work. The material excavated during the past month has been mainly boulders, loose rock interlaid with broken shale and clay, and some thin layers of hard sandstone, the most of which required blasting, but deposits near sound bedrock had to be carefully handled, much of it loosened with picks and wedges in order to avoid cracking the foundation."

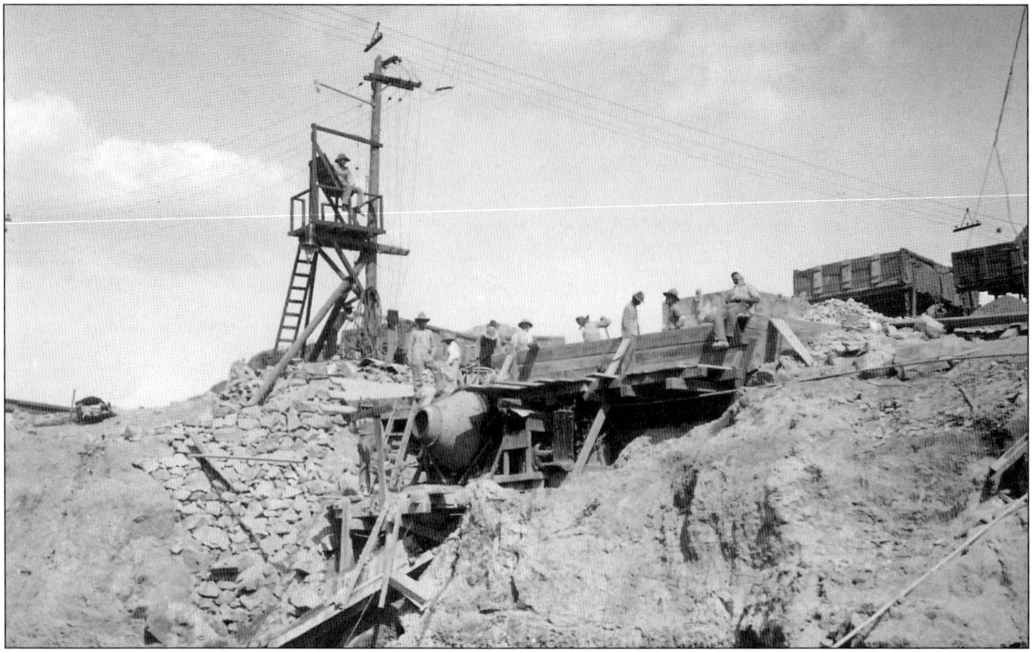

Three railroad tracks ran from the mixing plant at right angles to and beyond the cableways. This allowed one mixer to serve any cableway, and only one man was needed to handle the operation. It was noted that two men on each mixer gave the greatest economy when uniformity of the product was needed.

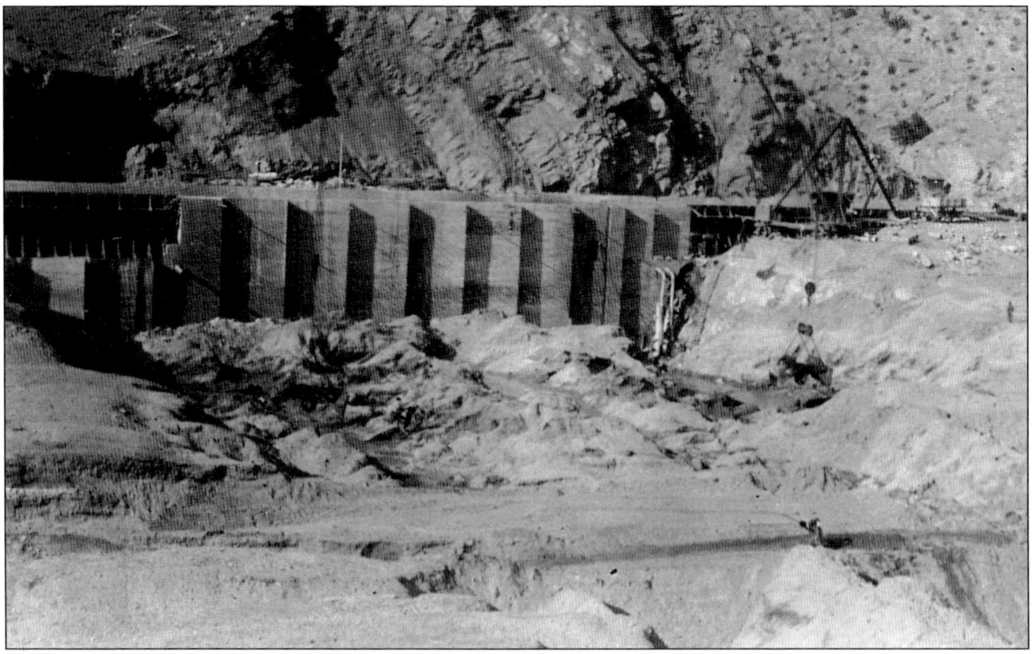

R. Swanzy spent Sunday looking over the Elephant Butte Dam project. He said it was a "stupendous piece of work," according to the *Deming Headlight* of September 13, 1912. For three years, lawyers and landowners had argued about the settlement being offered for the land needed for the dam. To see the actual construction come to fruition had to be as Swanzy described it—stupendous.

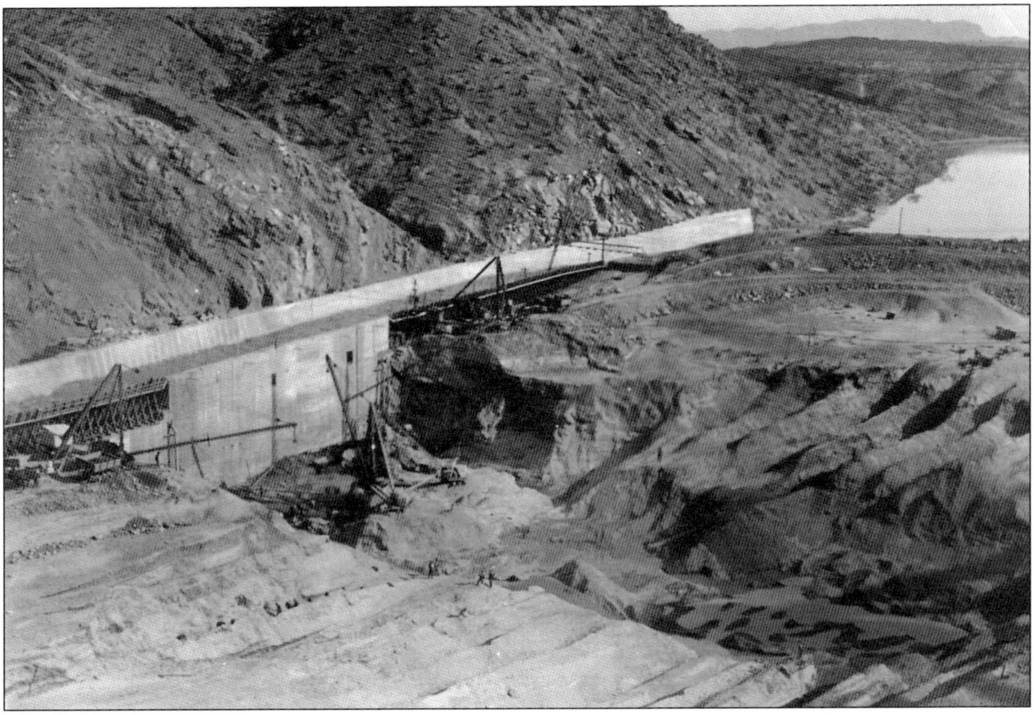

The *Waterloo Times Tribune* wrote on February 26, 1913, that "The unsung engineer in charge of this mighty work is E.H. Baldwin, a Cornell graduate who has been for many years identified with big government and municipal undertakings. Under his direction men become as Titans breaking the formation of (unknown) to shape themselves to their bidding. Already the river has been diverted into a narrow, deep sluiceway, leaving beside it 1,000 feet of dry river bottom through which the Mexican sand hogs and hard rock men are delving to bed rock. Huge cable towers have been reared on the canyon summits and tons of rock and cement are swung across the canyon on these cables as easily as your change comes back to you in the overhead cash carrier in the department store."

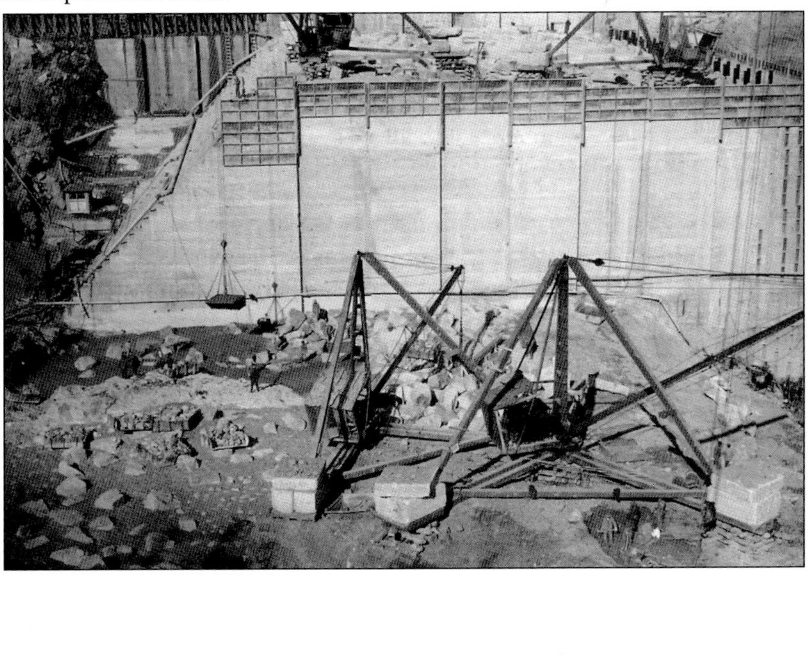

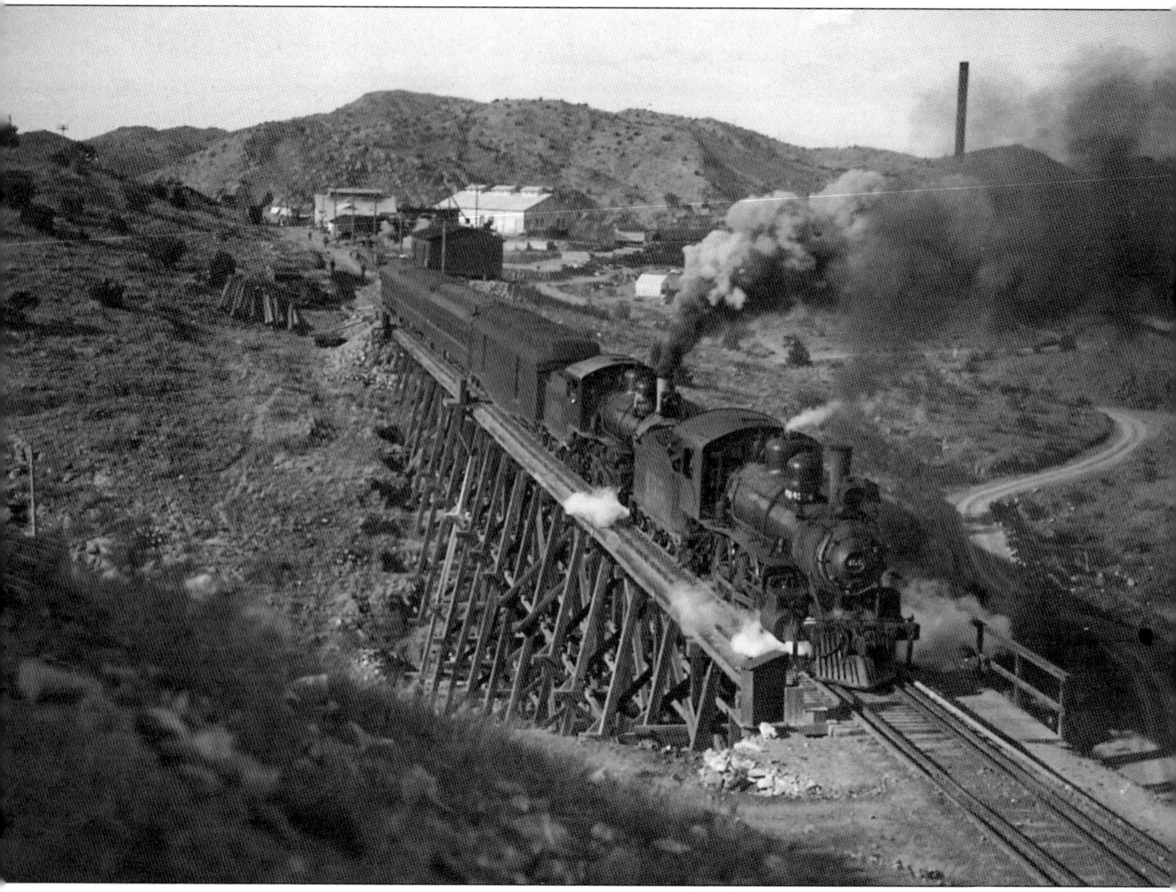

"El Pasoans who visited the Elephant Butte Dam work on Wednesday were moved by various opinions and degrees of satisfaction. The visiting St. Louis bankers, in whose honor the excursion to the damsite was made, expressed much enthusiasm over the work and the future prospects of the Rio Grande Valley as a result of the irrigation that it will afford them. Repeatedly they expressed themselves as very much impressed with the possibilities for the valley when the work is finished," wrote the *El Paso Herald* on December 5, 1912. The photograph above is the actual train that carried the El Paso group on their excursion, as documented in the *El Paso Herald*. It was noted that some of the El Pasoans "candidly and openly criticized the work as being slow, while others expressed satisfaction and ventured the opinion that it had progressed as rapidly as could be expected."

Men kept the mixers supplied with cement and crushed stone. All excavated material that was suitable for concrete was taken out by what was called a grab bucket and dumped in hoppers to be carried away to the cement storage bins and the rock-crushing and concrete-mixing plants. Pictured here is the continued work being done on the flume.

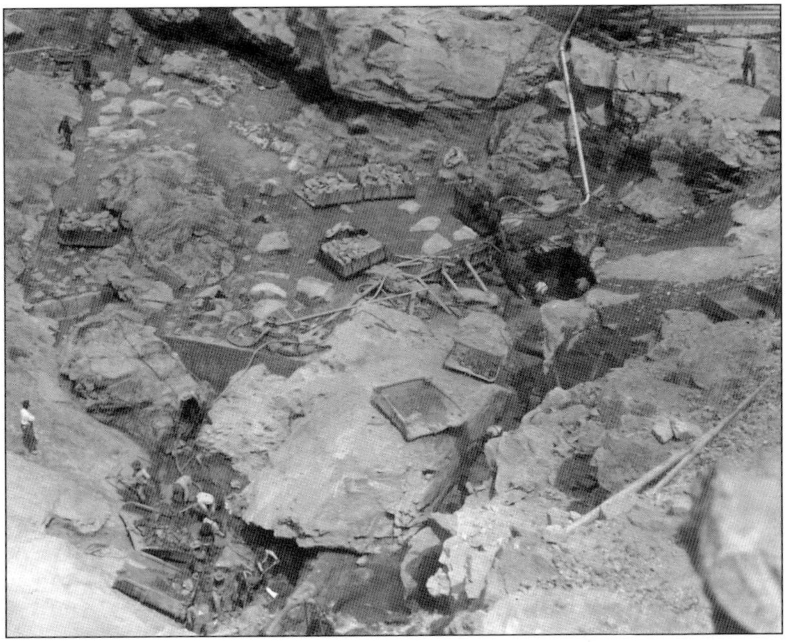

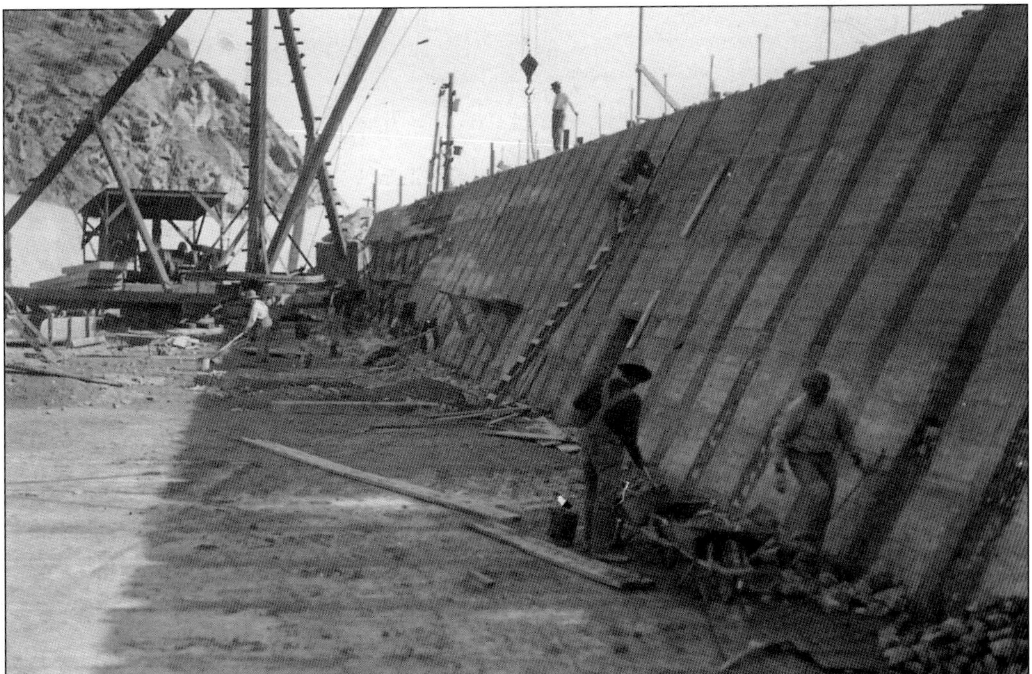

Engineers said that a very small amount of water was actually encountered during the excavation process, but pumping equipment was available if the need arose. Engineers were amazed that directly underneath the flume with the river flowing 4,000 second feet on top of the flume, there was not even a trace of dampness. This photograph provides a very clear view of the actual flume before the water was diverted into the structure.

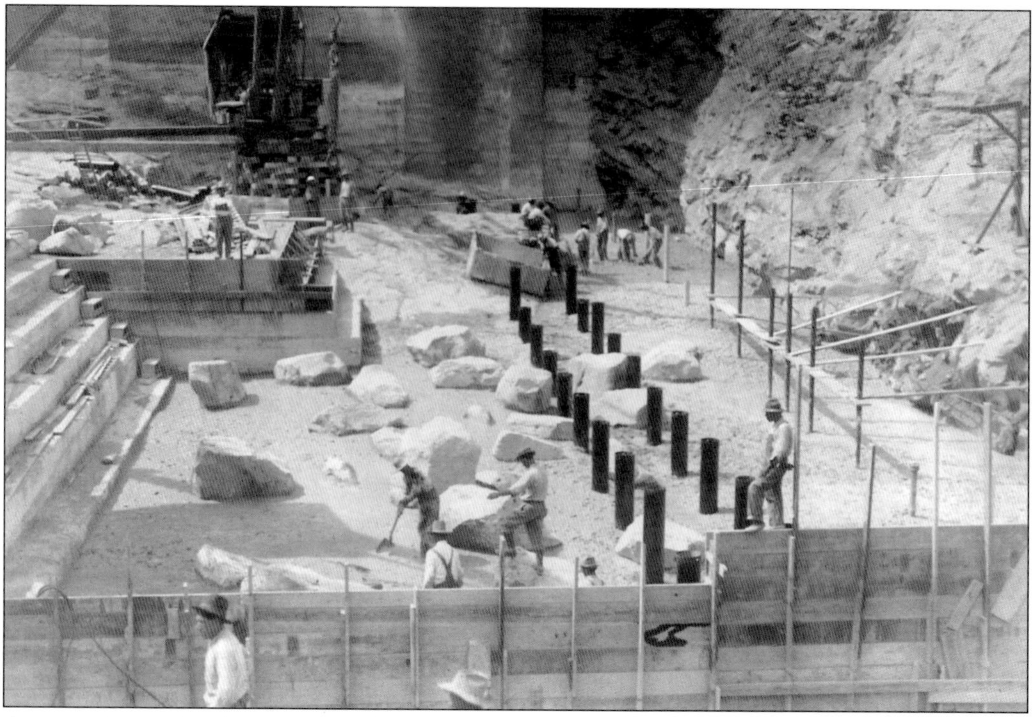

Competition at the damsite was a fairly common practice between different crews, called gangs. During the excavation, a man with a stopwatch plotted the time of each operation of the excavation and then followed up to determine how his crew could make the operation more efficient (faster). It was said that this fostered a "healthy rivalry" between the different gangs. The number of trips made by each cable and the yardage handled by each bucket was calculated and tabulated by a typewriter and posted daily for all the workers to see. It was said that the best day for grab buckets for three shifts with all cables in operation was a staggering 1,817 cubic yards. For the best run for three buckets in one shift, the winner hauled 722 cubic yards.

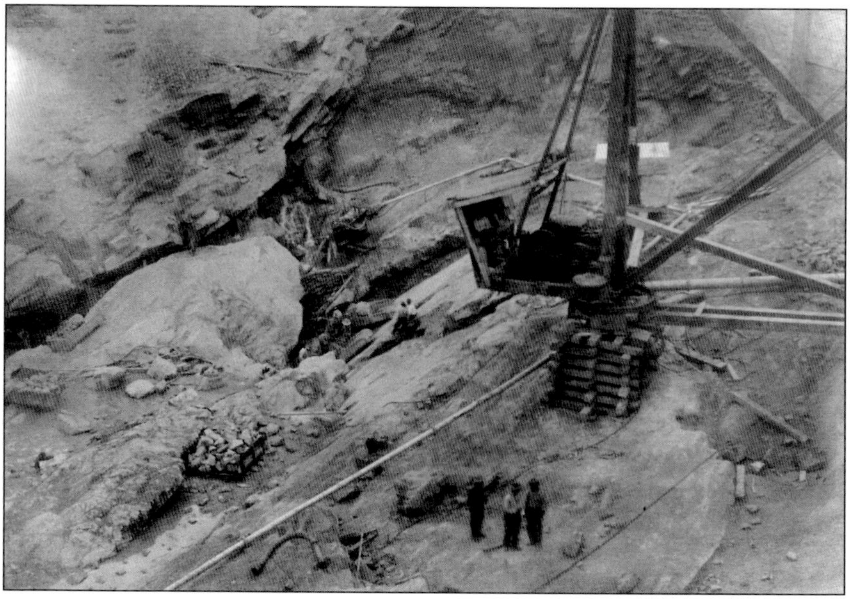

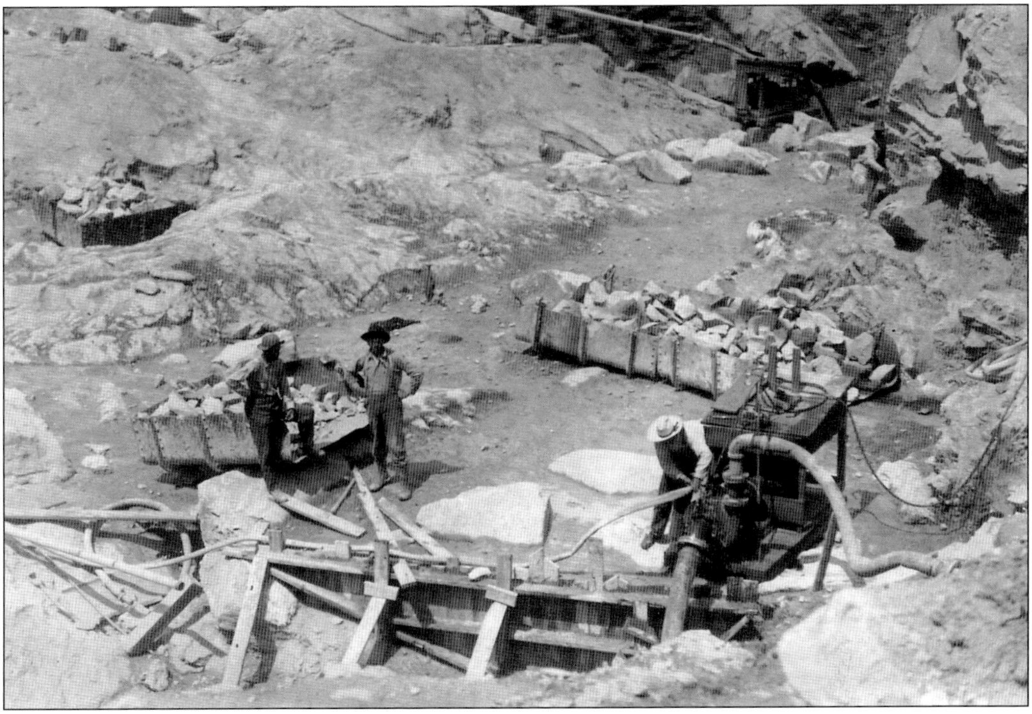

From its earliest construction, Elephant Butte Dam was being compared to Great Britian's Assouan (Aswan) Dam, built on the Nile River in 1905. According to the *Waterloo Times Tribune* of February 26, 1913, "The American public, used to tremendous government undertakings, scarcely is aware of the start of the work on the Elephant Butte dam which will supply for Rio Grande reclamation project in Texas." The paper went on to say, "No doubt you have read the wonderful accounts of the Assouan dam, the great British reclamation structure which is to impound the flood waters of the Nile and feed them down to thirsty African farms during dry seasons. No reclamation project has been so widely advertised as the one at Assouan. It is commonly regarded as the mightiest irrigation plan ever undertaken by man."

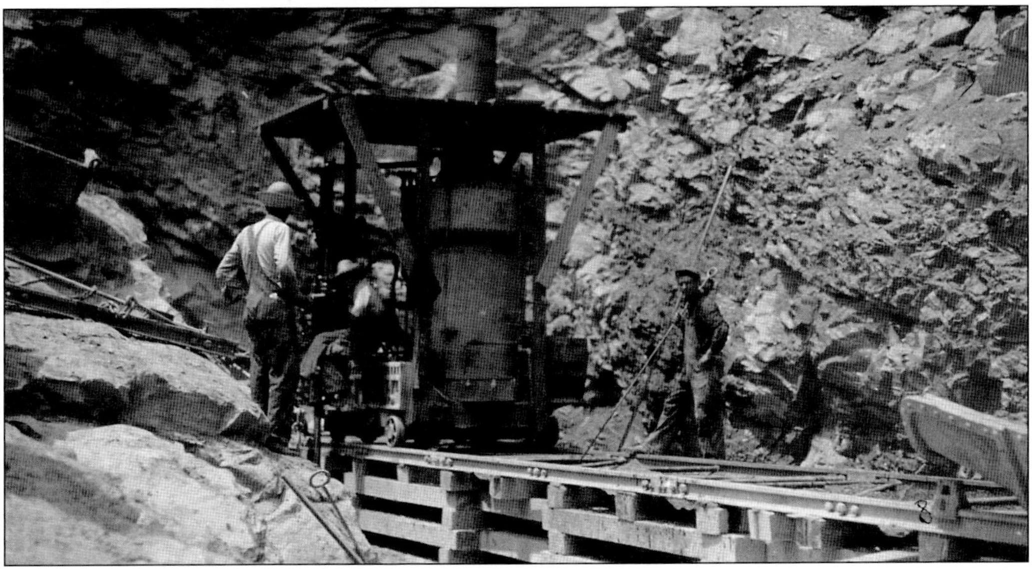

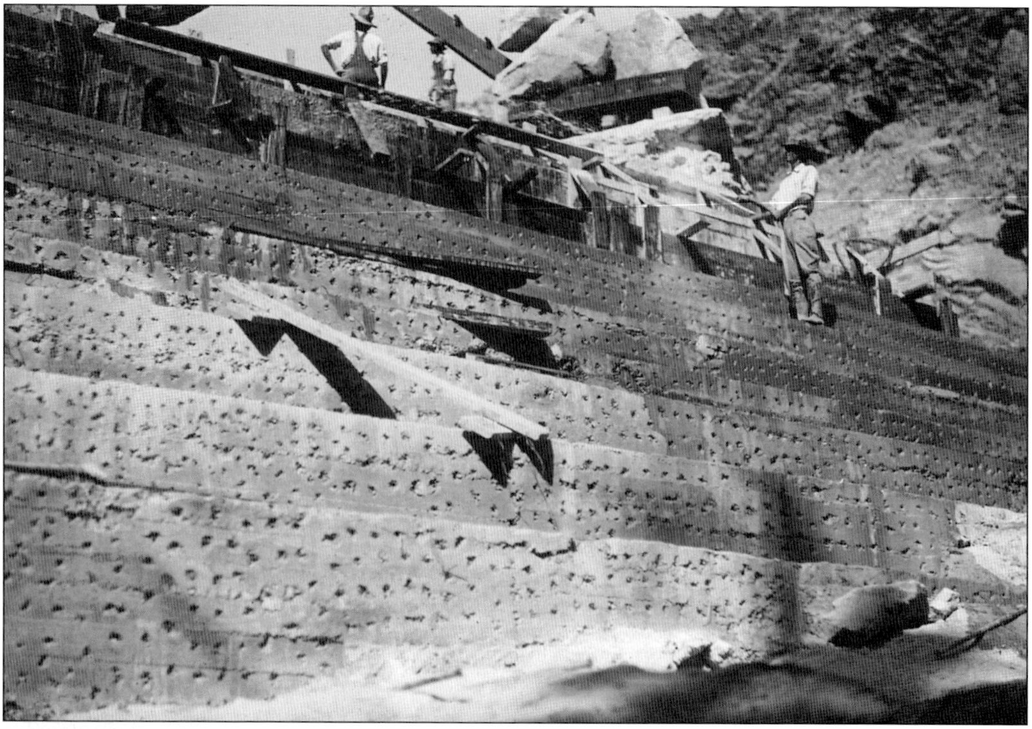

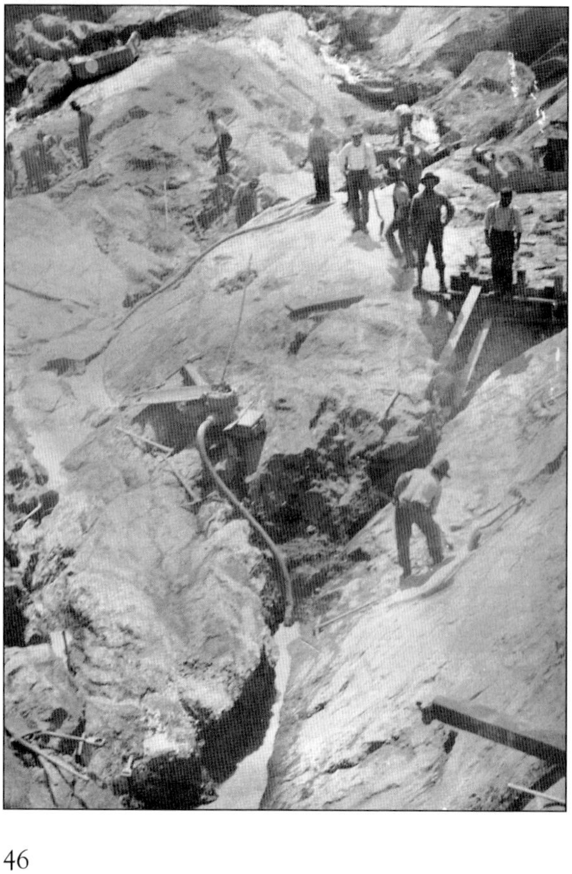

The *Abilene Semi Weekly Reporter* reported on former Abilene resident C.O. Cozort on February 2, 1912, who was now living in Las Cruces, New Mexico. Cozort said that Elephant Butte Dam would irrigate "millions of acres of lands," making New Mexico "blossom like a rose." The photograph shows the steps to the cut-off trench for the flume in August 1912.

According to territorial newspapers, the first concrete of the dam structure was poured on June 3, 1913. From that day on, the process for all operations of the construction of the dam were timed in order to account for the greatest pouring of concrete in the shortest amount of time.

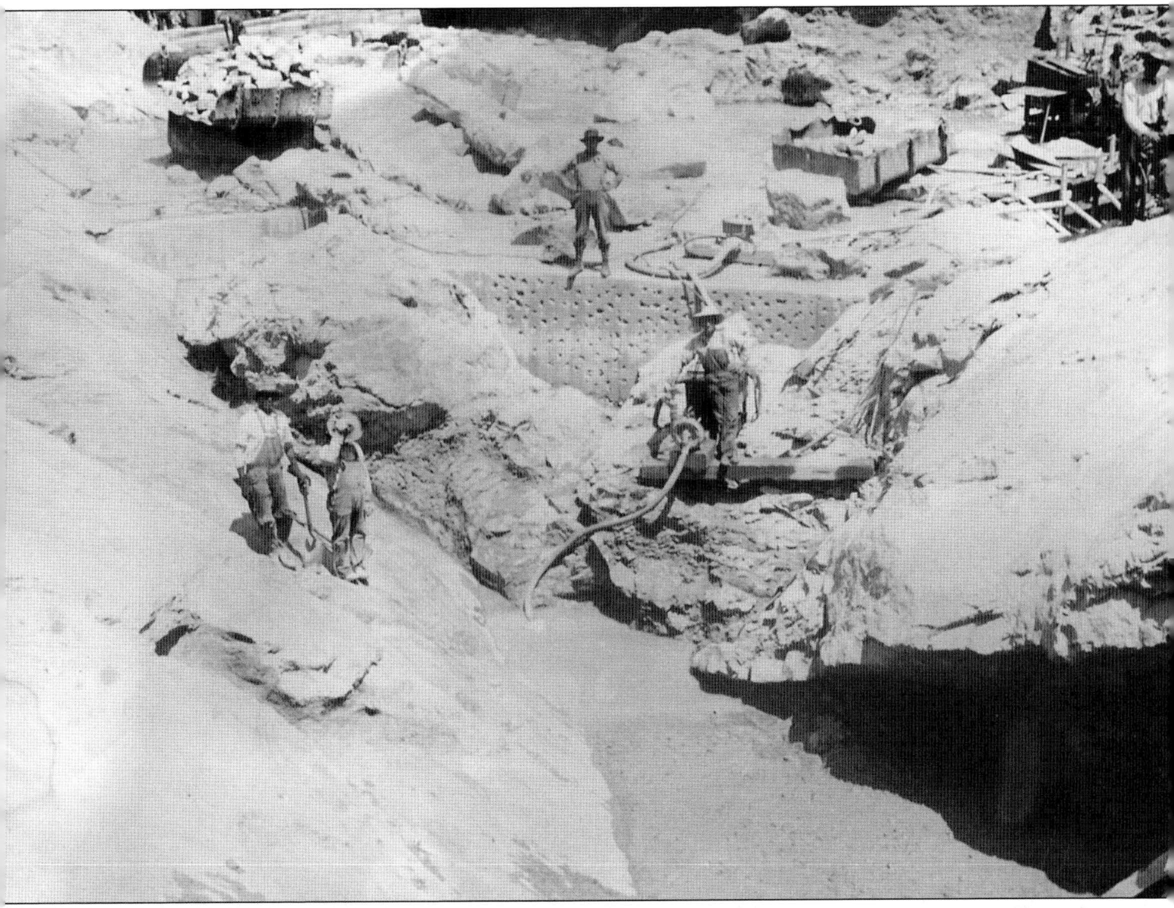

During the excavation, a total of 204,000 cubic yards of material for use in masonry was collected and stored. The holes that were drilled were for "light charges," which used blasting powder. As the sand and gravel in the damsite was removed to a certain depth, drag scrapers worked in pairs run by two-drum electric hoists located opposite each other. The drum on one hoist pulled the loaded scraper forward while the back line was reeling off from the drum of the second scrapper. Whatever material was not suitable for concrete aggregate would be dumped into the flume for the river to carry out or placed into a dump car and taken by a locomotive and dumped onto the cofferdam. The material that was deemed suitable for use was taken by grab buckets to the upper railroad and dumped into storage hoppers.

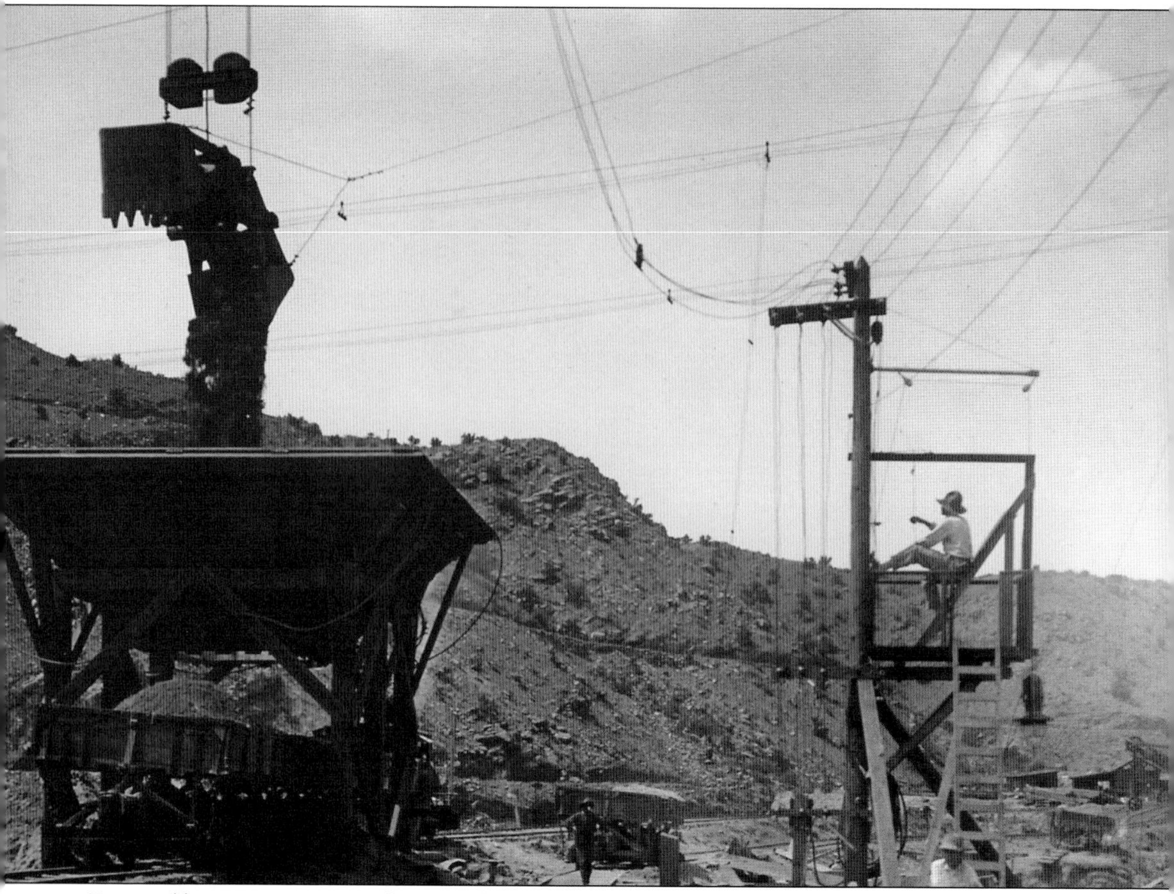

Pictured here is an operator directing his load using the elaborate system of Lidgerwood cableways. The power plant at the damsite furnished 60-cycle currents at 2,200 volts, and the current was distributed at this voltage and used directly without transformation by the cableways and some of the large motors on the damsite. Lidgerwood Manufacturing Company was famous worldwide for their hoists and boilers, and the company was located on Ferris Street in Red Hook, Brooklyn, from 1882 until 1927. They also had offices at 96 Liberty Street in Manhattan. It was from this office in 1900 that they advertised the Lidgerwood cableway as a hoisting and conveyance device. Interestingly, the Lidgerwood cableway was installed on the USS *Collier Marcellus* in the fall of 1899 and tested on the high seas during November and December of that year. It was utilized to transfer ammunition, provisions, supplies, and coal from one ship to another during what was considered to be "moderate sea and weather."

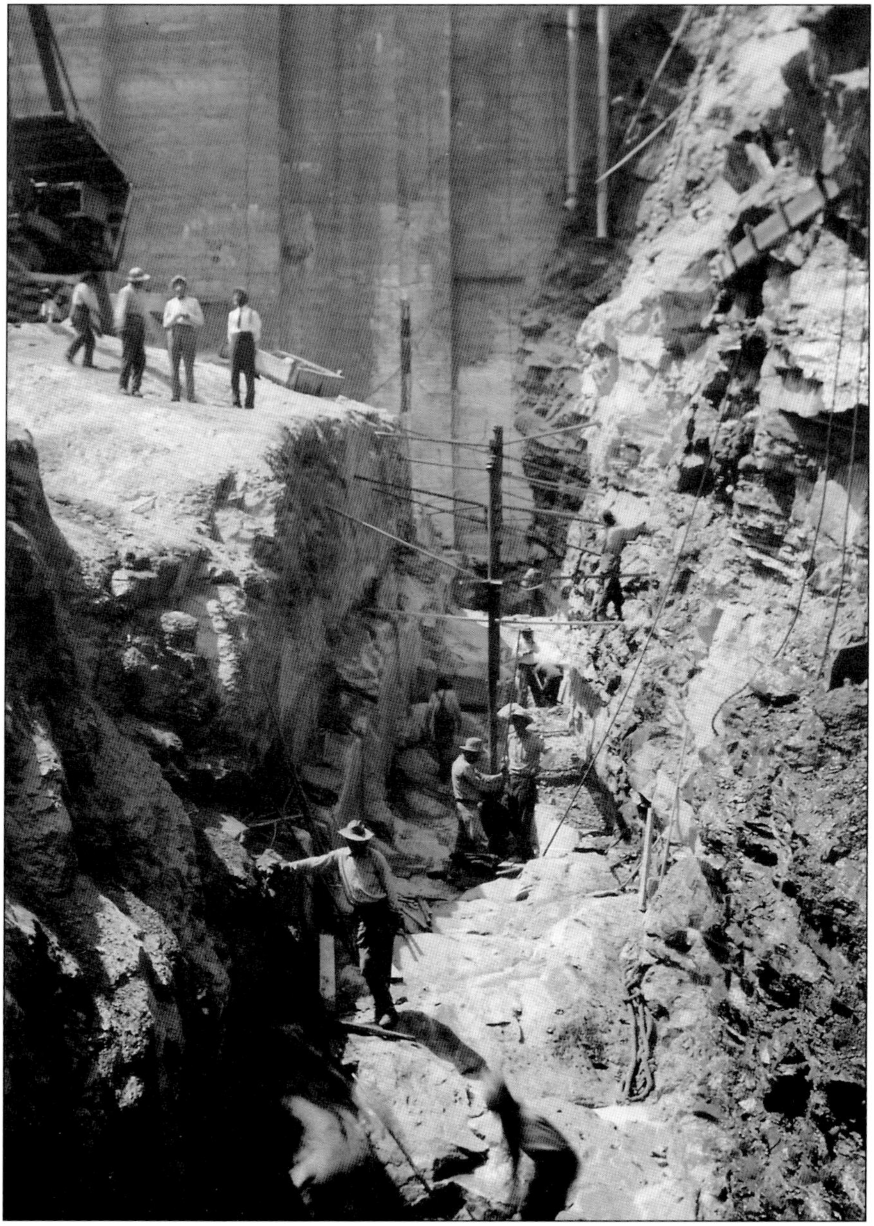

This photograph shows the cut-off trench being made in the bedrock by the channeling machine, forming the heel of the dam. The cut-off trench averaged 10 feet in width and 15 feet in depth, and the upstream side had a larger portion that was 50 to 100 feet deep. The downstream side had to be cut by a channeling machine, and the opposite side had to be drilled with holes that were very close together. This would allow the material to be loosened using light charge of powder. The cut-off trench was harder to excavate because of cramped working quarters and the constant threat of falling rock. So how did the engineers decide that this was the perfect spot for the dam? In the early 1900s, when the project was still being hotly debated, 37 holes, at an average depth of 75 feet, were bored by a diamond drill, penetrating rock to an average depth of 23.5 feet. The information that came from these borings provided the engineers with a sample and analysis the rock foundation.

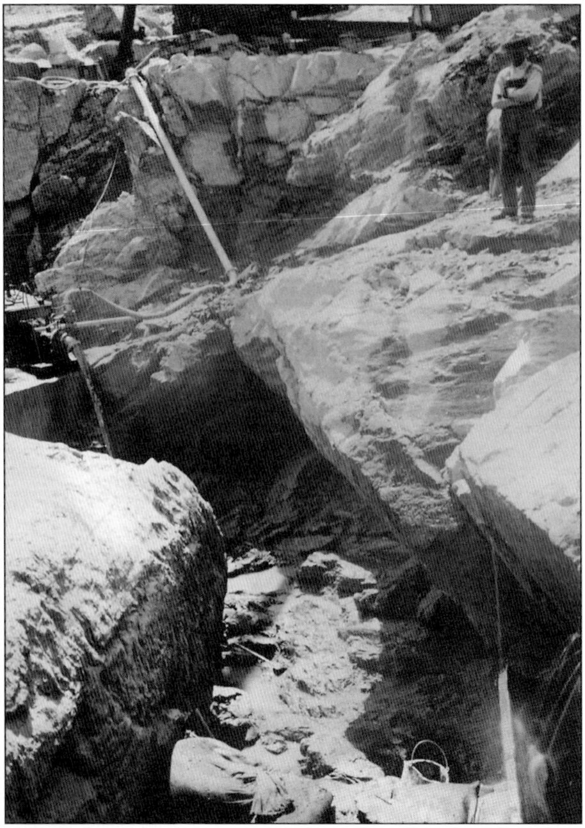

The photograph on the left is titled "Hogan's Alley" and dated July 1913. The bottom image was part of the construction of the flume. It was also in 1913 that there was some controversy at the damsite. Mexico and the United States were arguing over a G.M. Putnam, commander of the guards at the dam. The New Mexico courts had acquitted Putnam of the killing of a Mexican boy at the damsite in 1912, although he had admitted to killing the boy. No specifics were given. In 1913, Putnam crossed over into Mexico on a sightseeing trip and was recognized by somebody who alerted the boy's father, a noncommissioned officer in the Juarez federal army. The father then alerted the Mexican authorities, and Putnam was thrown into a Juarez jail.

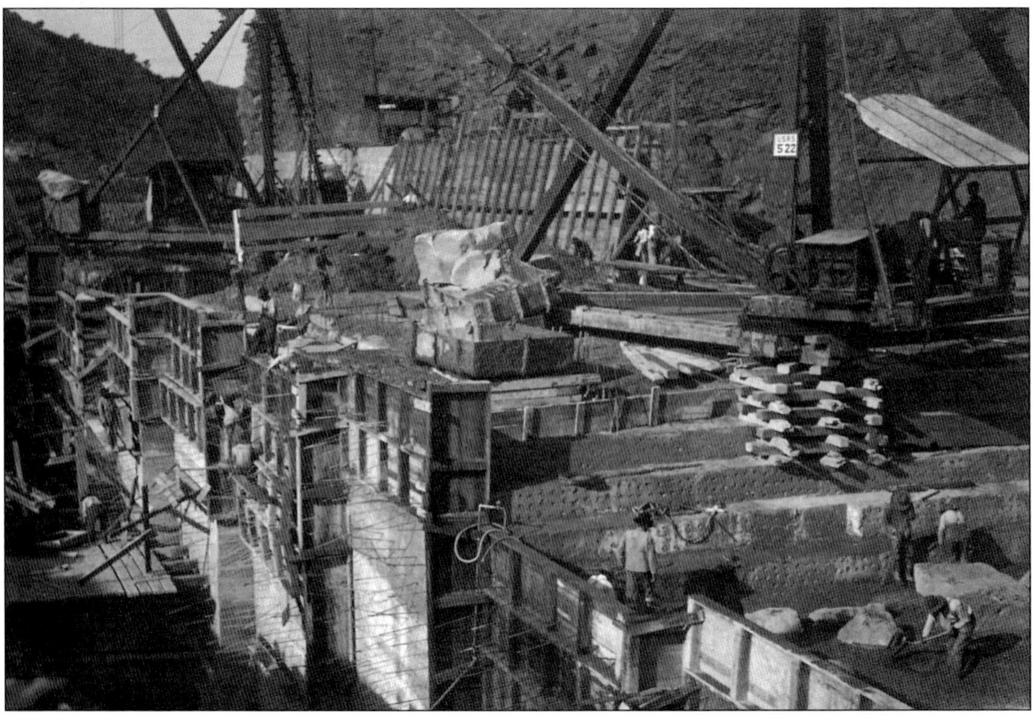

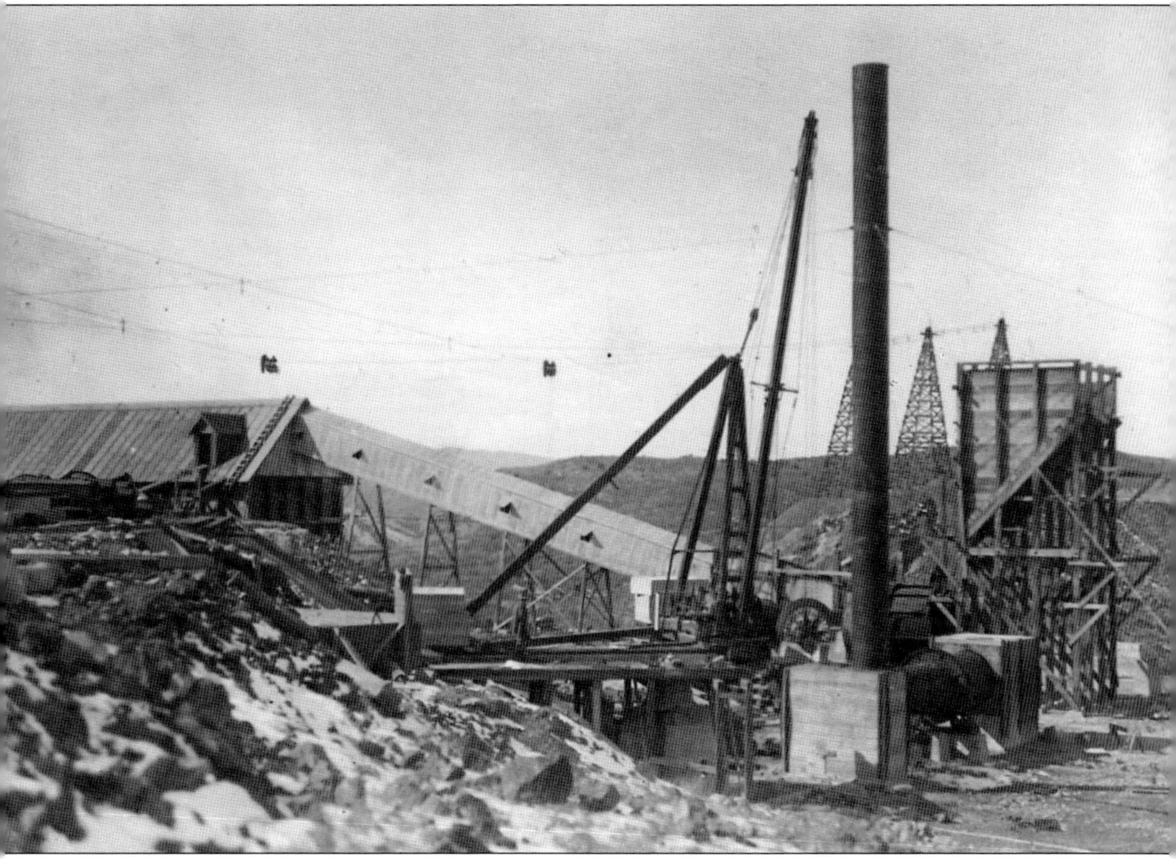

Pictured here is part of the sand-cement storage bins that were located close to the concrete mixing plant. R.R. Coghlan, who was in charge of the cement plant on the Roosevelt Dam, had designed the sand-cement plant at the Elephant Butte damsite. The plant facility contained a gyratory crusher, rotary dryer, ball mill, mixer for mixing the cement and ground rock, and four tube mills for reducing the mixture to a fine powder. The total cost of the mixing plant, which included the machinery, storage bins, elevators, and concrete-haulage system was $57,800. The crusher plant had two No. 7.5 Symons gyratory crushers working at a crushing capacity of 100 to 150 tons of material per hour. Each of the crushers had an elevator that delivered the product into the storage bins. The crushers were so deadly that a house was built over each crusher to protect the men operating them.

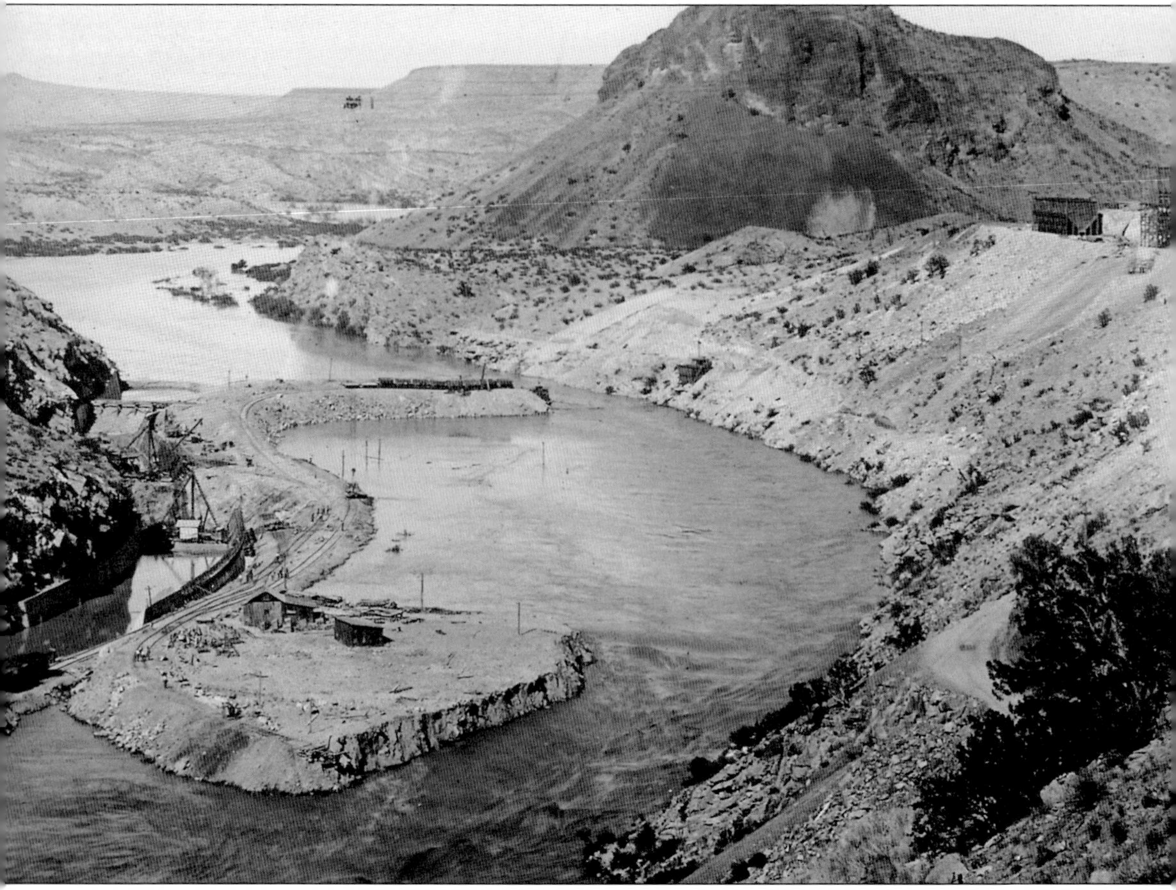

It would not be fair to leave the story of Jose Sanchez behind. The photograph is labeled "high water," and that is exactly what happened to poor Sanchez. In October 1911, Jose Sanchez, an employee from the Elephant Butte construction gang, was walking with a companion along the bank of the Rio Grande. The river was higher than usual, just as it was in the image above. The high water had significantly compromised the river bank and gave way under Sanchez's footing. Unfortunately, Sanchez was not able to regain his footing and consequently drowned in the cold water of the Rio Grande. (It is presumed that his friend could not swim either.) More than likely a current swept his body away, and it was not until November of that same year that Sanchez's body was found in a stagnant pool of water below Palomas Hot Springs, New Mexico, just six miles from where he had drowned. His body was so badly decomposed that he was barely recognizable. His wife, however, had been called to the scene and confirmed it was her husband. Sanchez had previously lived in Las Cruces, New Mexico.

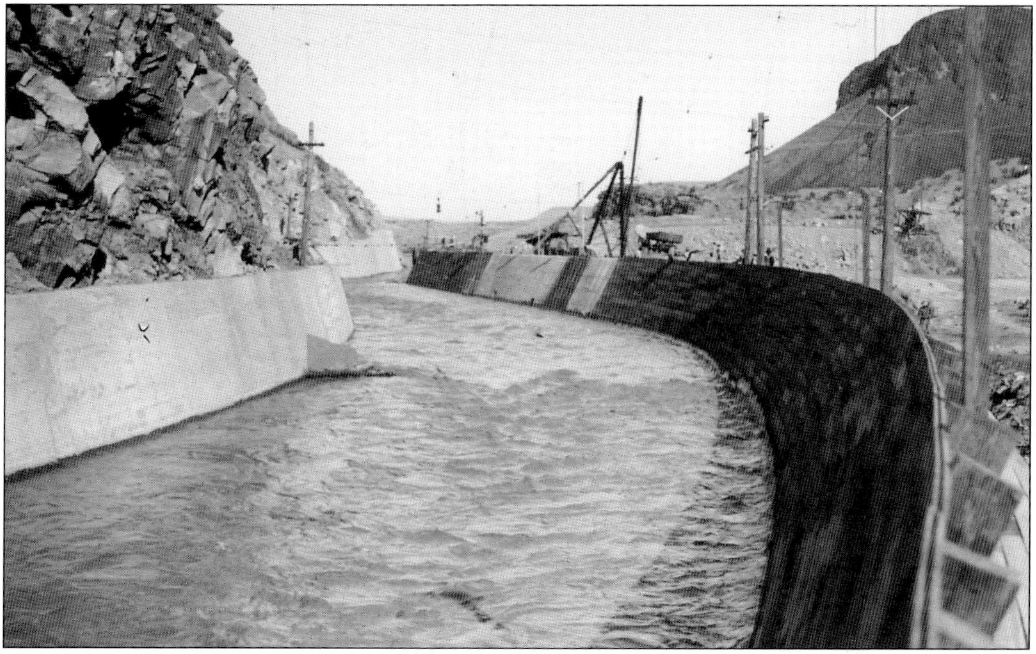

The photograph above shows the waters of the Rio Grande flowing through the flume in November 1912. The flume was 1,200 feet long, averaged 50 feet in width, and was 16 feet deep. As the *Waterloo Times Tribune* had forecasted on February 26, 1913, "The Rio Grande is variable in its flood seasons. During many months of each year it is so low that children can wade across it. Sometimes it goes two years without a flood." The engineers reported that they had constructed the flume to handle any seasonal flooding that might occur on the Rio Grande during the construction. The image below shows no water at the flume's outtake. Interestingly, an anonymous quote has been used to describe the Rio Grande as "too thin to plow and too thick to drink."

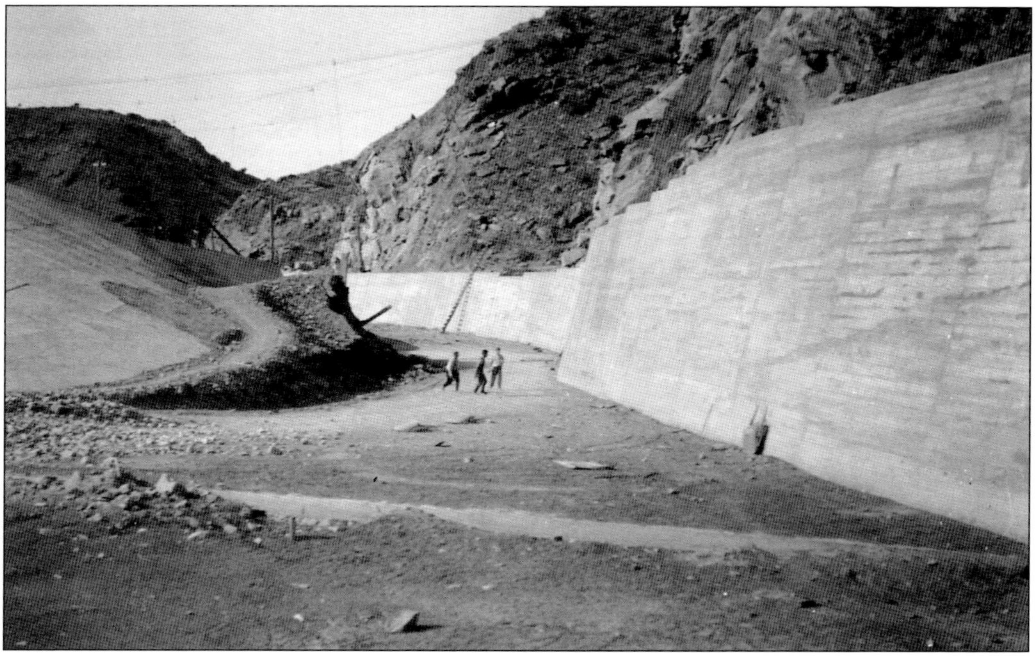

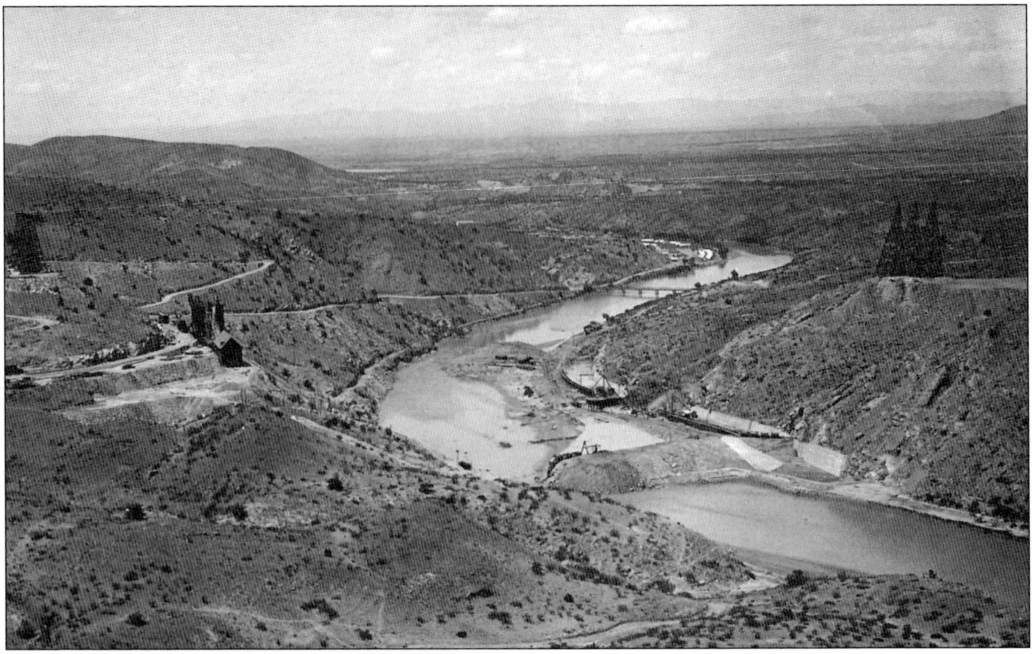

These two photographs bring to mind Ecclesiastes 3:1, which says, "To everything there is a season." The seasons for the flume had passed. The Rio Grande had been diverted. The December 25, 1914, edition of the *Rio Grande Republic* wrote, "When completed, the big dam will be 1,200 feet long on top at an elevation of 4,414 feet, which is 200 feet above the elevation of the original river surface. The maximum height from the deepest excavation to the top of the parapet wall will be 304.5 feet. The maximum base width is 215 feet. It will form a lake in the Rio Grande valley averaging 134 miles wide, 66 feet deep, and with a shore line of 200 miles. The lake will cover 42,000 acres and contain 862,200,000 gallons of water."

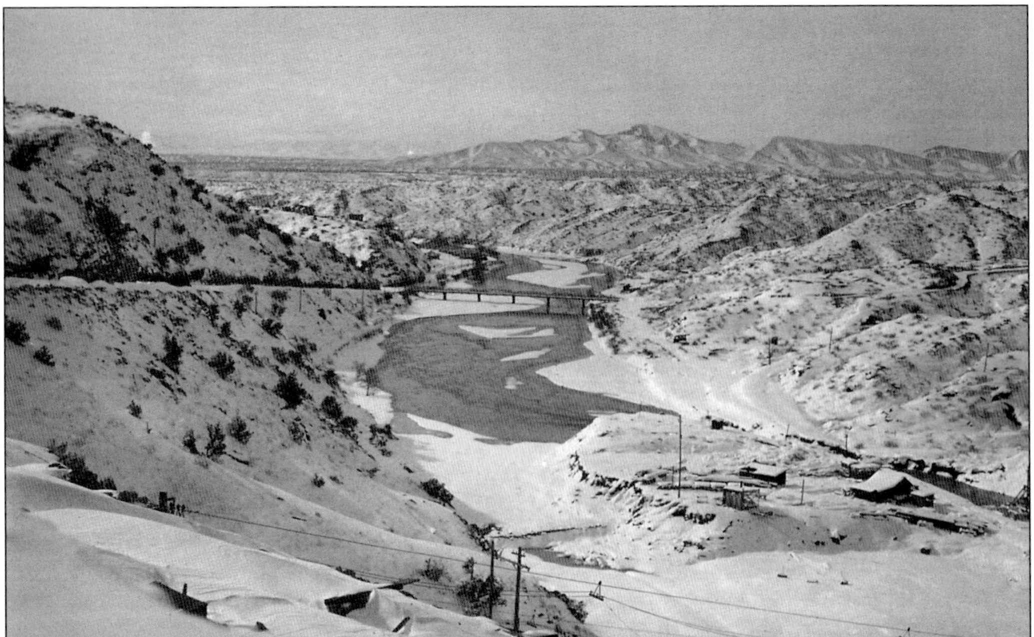

Three

Living in the Camps

The Segregation

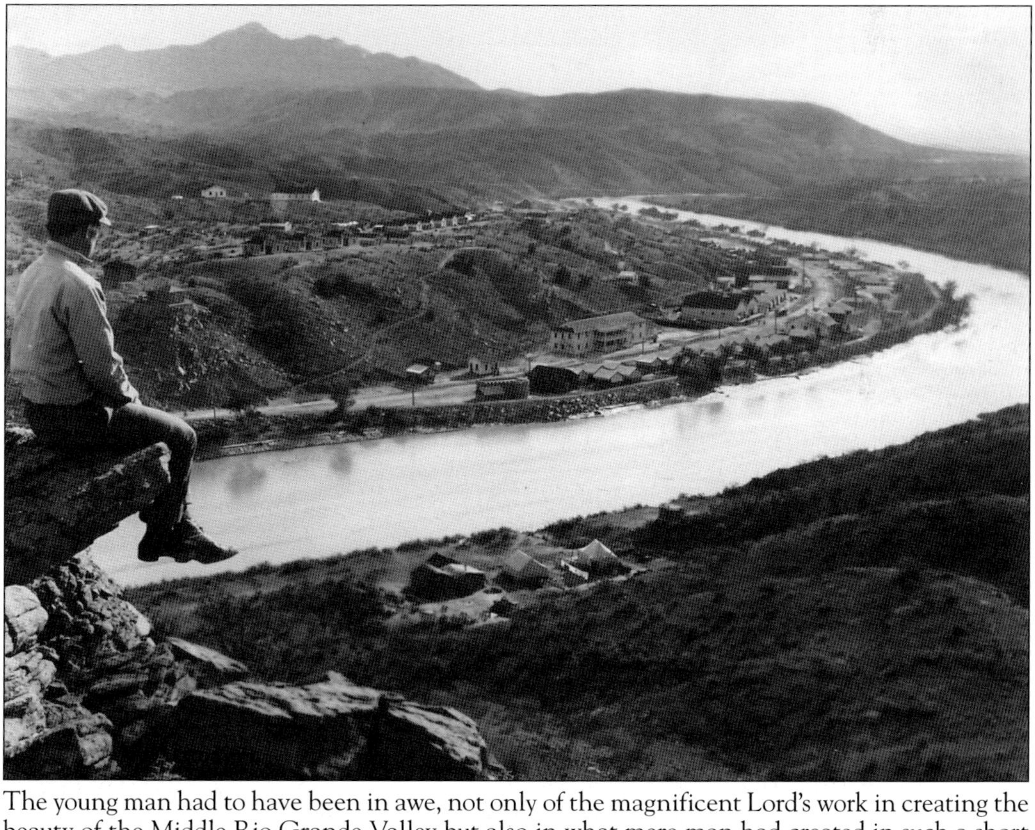

The young man had to have been in awe, not only of the magnificent Lord's work in creating the beauty of the Middle Rio Grande Valley but also in what mere man had created in such a short time at the townsite of Elephant Butte Dam. From Psalm 24:1, "The earth is the Lord's, and the fullness thereof; the world, and they that dwell therein."

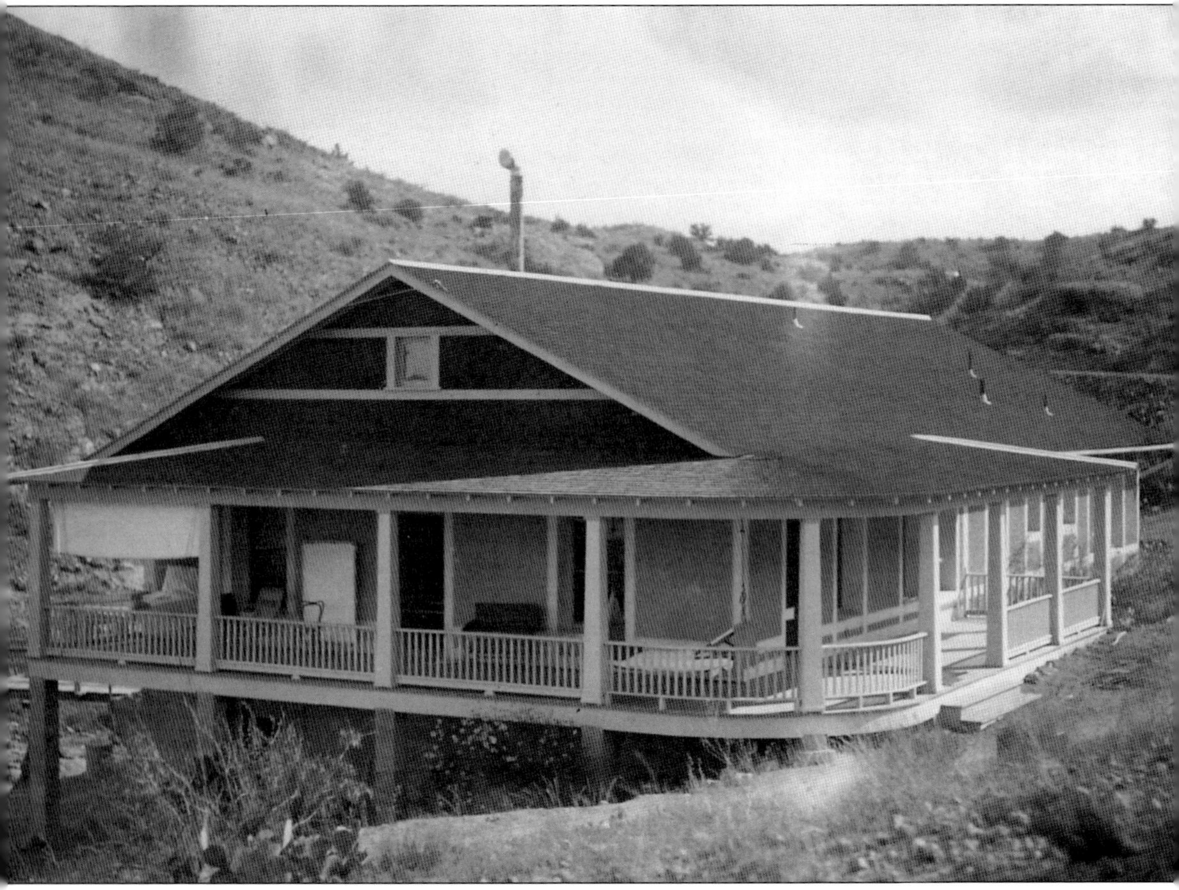

Dr. Dale Graham, camp physician at the damsite hospital above, had originally brought his family out west to Artesia, New Mexico, because of his wife, Madge, who had tuberculosis. A week after arriving at Elephant Butte, Madge died. Despite his obvious grief, Dr. Graham continued as the camp physician. In 1914, Dr. Graham wrote about life in the camps in *Engineering News*, providing a detailed firsthand look at the lifestyle of the dam employees. Dr. Graham was assisted by a nurse, an orderly, and a housekeeper/cook. The hospital had an operating room, dispensary for medications and supplies, and a consulting surgeon who made regular visits. It was noted that stretchers, crutches, and first-aid kits were readily available at the worksites for emergencies. Dr. Graham was quick to note that drugs and hospital supplies had been bought in large supplies. There was no short supply of peroxide, cotton, and bandages.

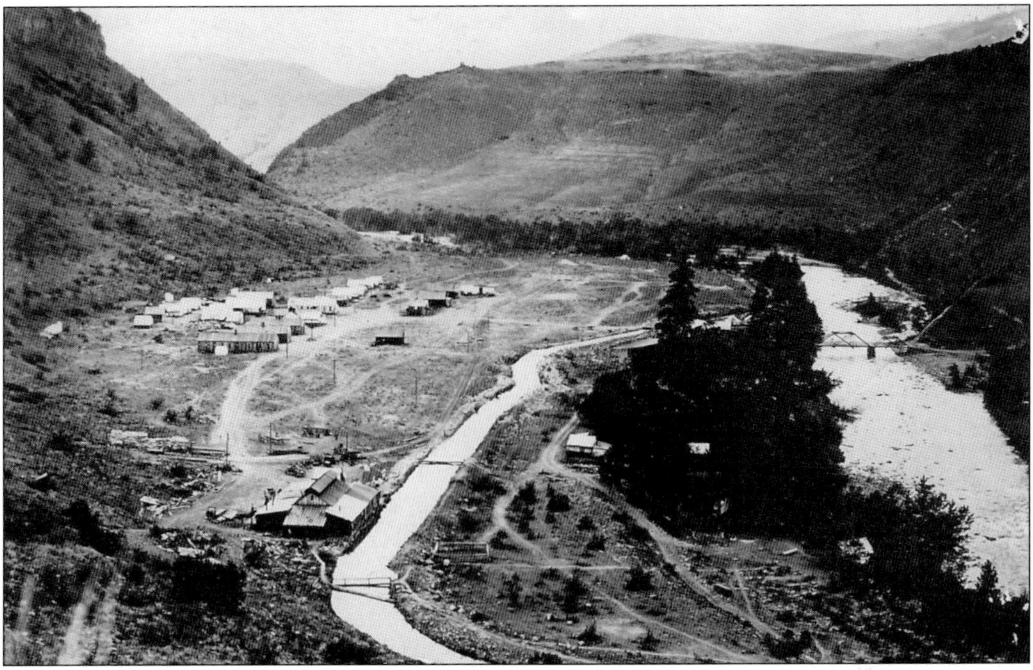

The photograph above is identified as "Bird's eye of Camp One," indicating that this was probably the first camp established at the damsite. Laverne Charles, whose various positions at the damsite included serving as acting chief of construction, wrote, "Our house is down the road below where the cross is." Charles's images have provided a firsthand glimpse into the construction of Elephant Butte Dam.

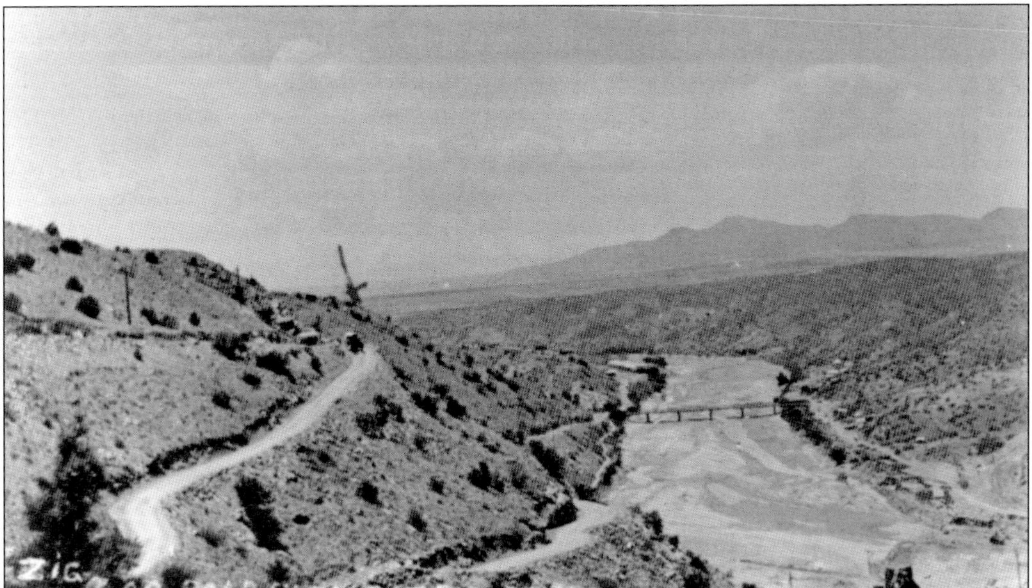

Pictured here is Zig Zag Road, which led to nearby Palomas Springs, New Mexico, where residents bragged that drinking or soaking in the natural hot springs would cure anything from arthritis to tuberculosis. Camp residents were discouraged from drinking the hot-spring water from Las Palomas, as measles and diphtheria were rampant at that time.

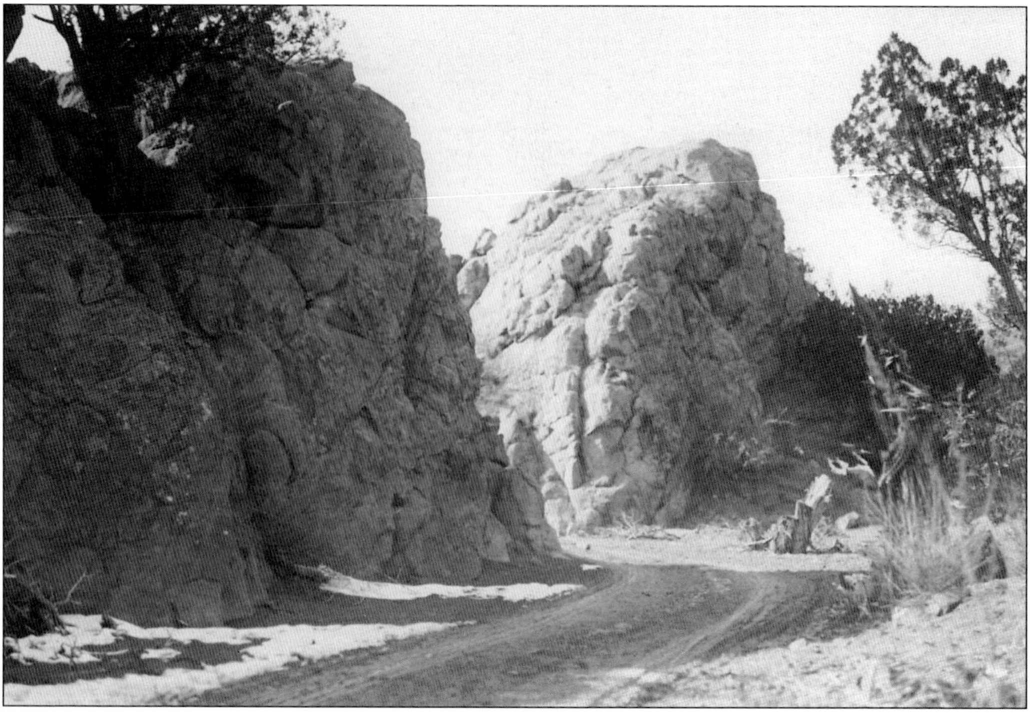

The photograph is labeled "The gap on road to Palomas Springs." Palomas Springs was close enough for camp residents to shop, soak in the springs, or visit the local bar. Palomas Springs would later be renamed Hot Springs and then Truth or Consequences. See Images of America: *Truth or Consequences* for the tantalizing tale.

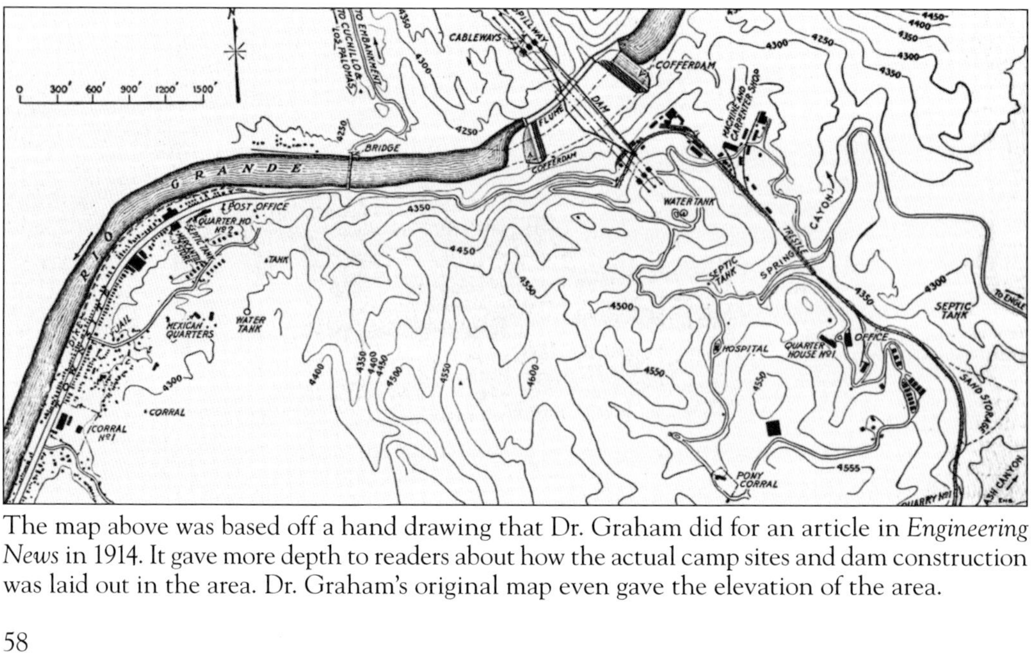

The map above was based off a hand drawing that Dr. Graham did for an article in *Engineering News* in 1914. It gave more depth to readers about how the actual camp sites and dam construction was laid out in the area. Dr. Graham's original map even gave the elevation of the area.

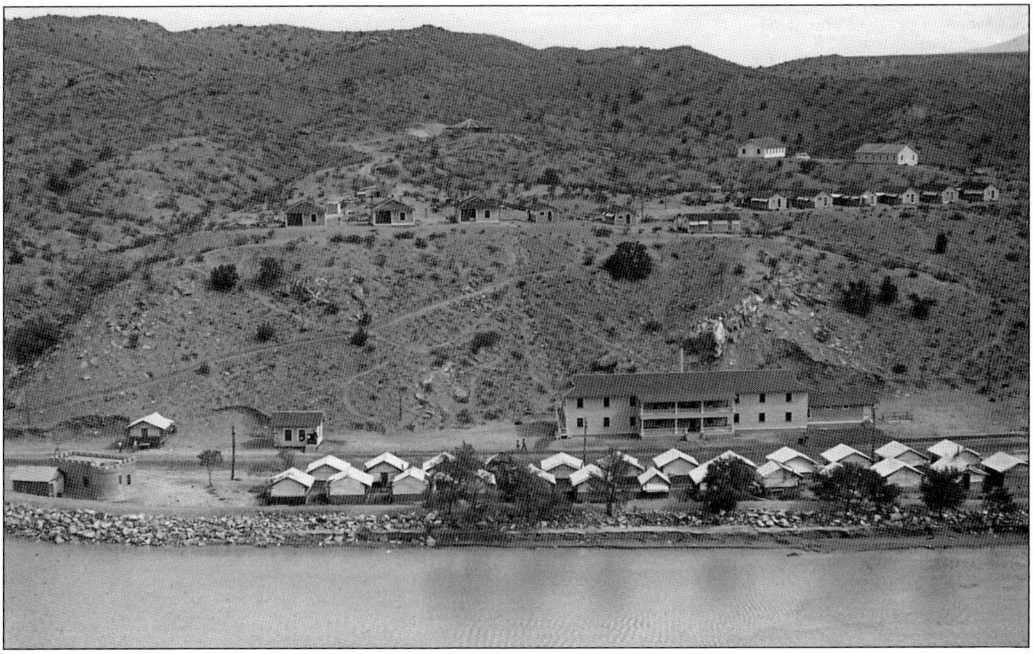

The Lower Town, seen above, was home to the foreman, mechanics, and laborers. Dr. Graham noted that the Lower Town was further divided into both an American quarter and a Mexican quarter. The Lower Town contained a commissary, mess hall, mechanic's quarters, bunkhouses for the single men, schoolhouses, a moving-picture theater, churches, and cottages. The two-story building, 26 by 150 feet, seen below and in the photograph above, was the mechanics' quarters. The 32 rooms were said to be capable of accommodating about 66 men. Part of the building was equipped with 12 "shower baths" outfitted with urinals and lavatories. The buildings had steam heat and electric lights. As mechanics would be considered skilled laborers, their jobs were held by the so-called Americans. Dr. Graham's map noted the segregation of Americans and Mexicans.

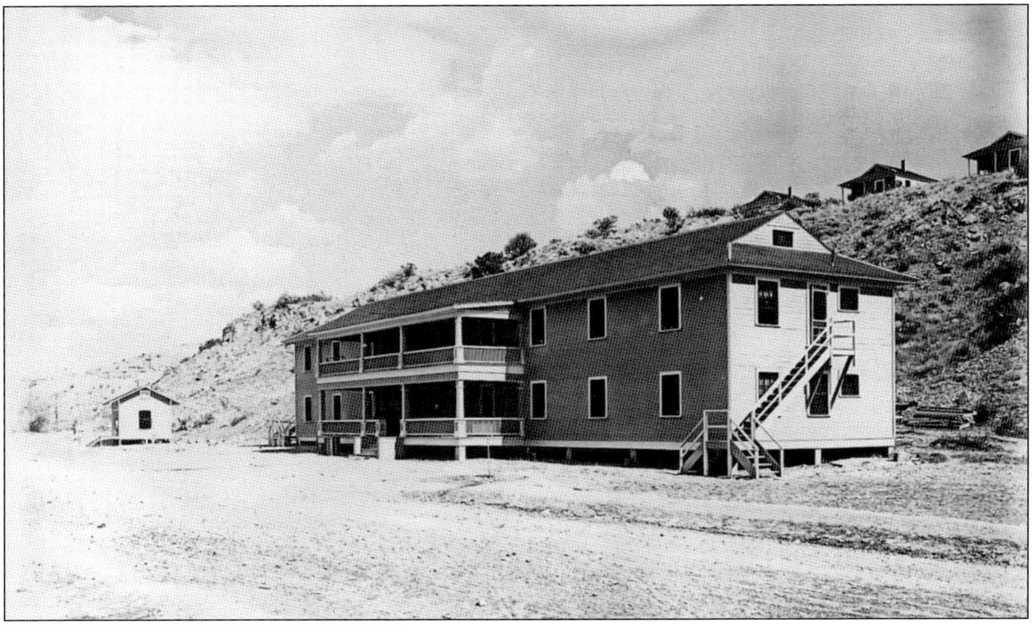

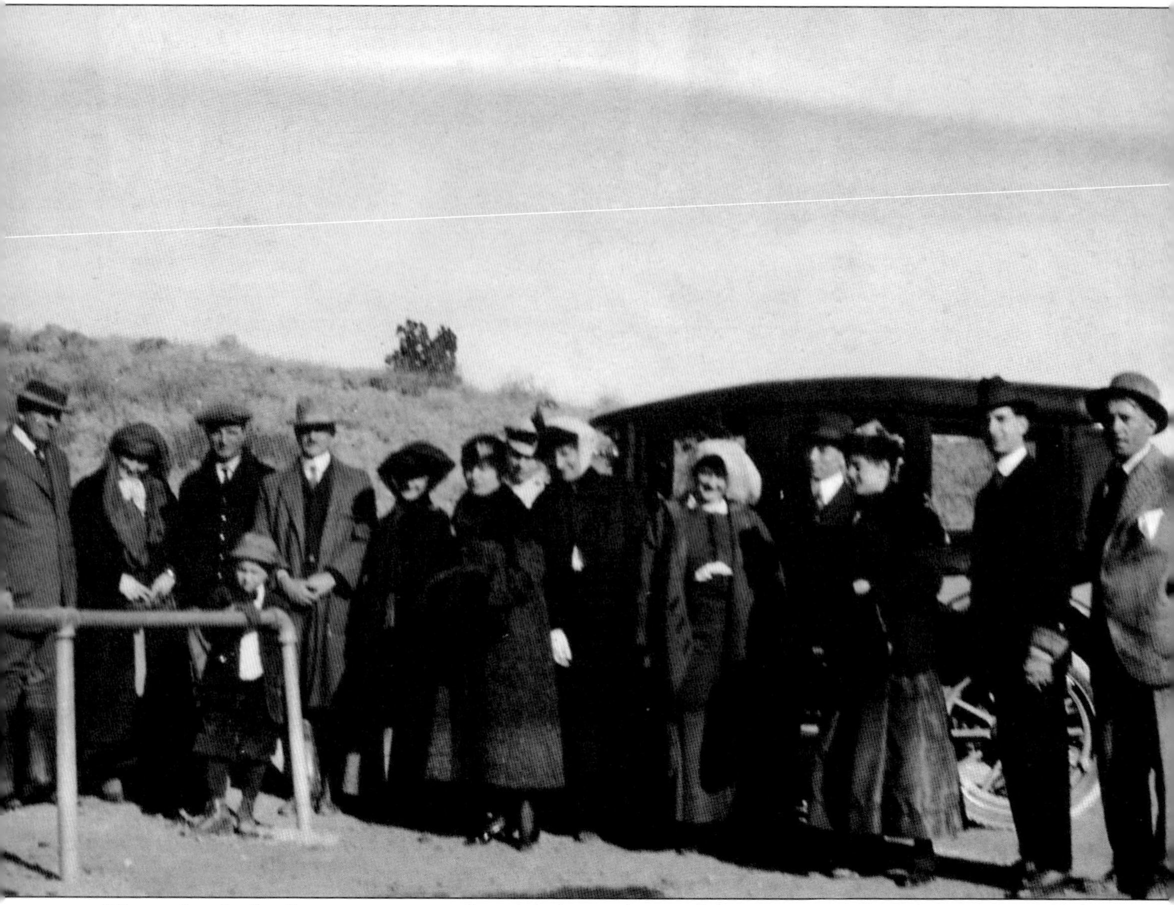

The individuals above were more than likely residents of the Upper Town. There was an obvious segregation between workers and those who lived up on the hill. This photograph came as part of the Laverne Charles Collection, and only five last names were recorded on the back of the image: McMarry, Charles, Pershing, Goldsture, and Frasure. Charles had several jobs during the Elephant Butte Dam era. He was listed during the Elephant Butte Dam construction as a construction engineer, assistant engineer, and acting chief of construction. The Colorado School of Mines listed him in association with the North Yakima Bridge in Washington State. Charles was born in Colorado in about 1878, and in 1940 he was living in Minneapolis, Minnesota, with his wife, Susan. He listed his occupation as civil engineer and listed his education level as having completed his fourth year in college. At the time of the 1940 census, he was still working for the government.

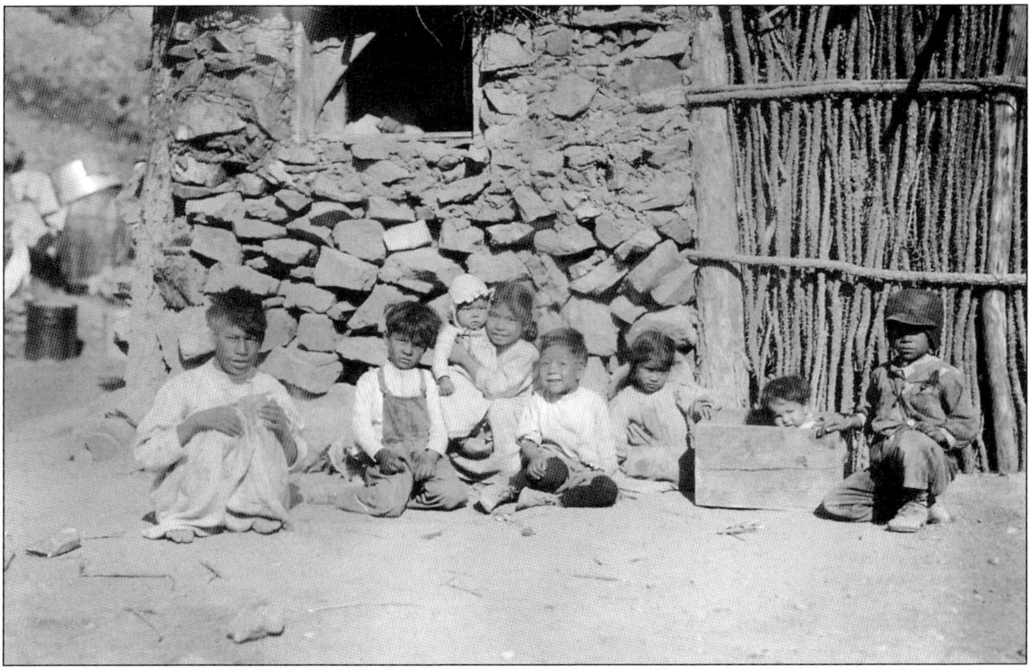

"Typical young Mexican citizens" was written on the upper photograph, and the image below had the following inscription: "Mexican residence of the better class." Workers that were white were called Americans, and those of Hispanic heritage were called Mexicans. Typical of the times, the Mexican and American laborers were separated in their living quarters and their lives. A Mexican boardinghouse was run by a Mexican family, with the Reclamation Service providing the building and the cooking stove. The family provided meals at a lower rate than the mess hall provided, and it was said that the food was more agreeable for the Mexican workers than the meals at the American mess hall. Allegedly, most of the skilled and supervisory roles were obtained by Americans. Mexicans, who made up two-thirds of the population, worked at the unskilled jobs.

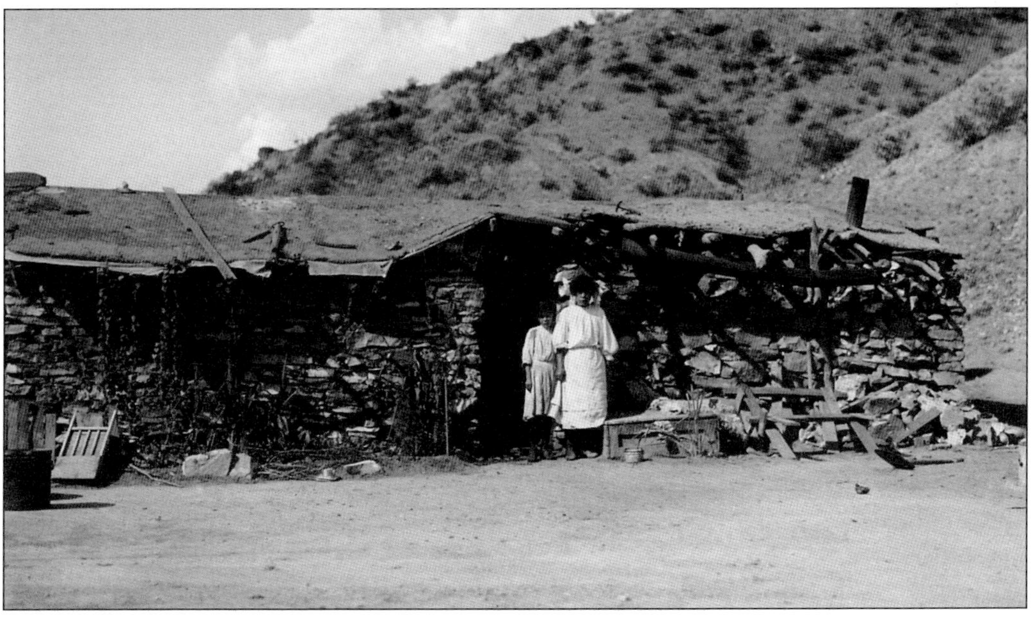

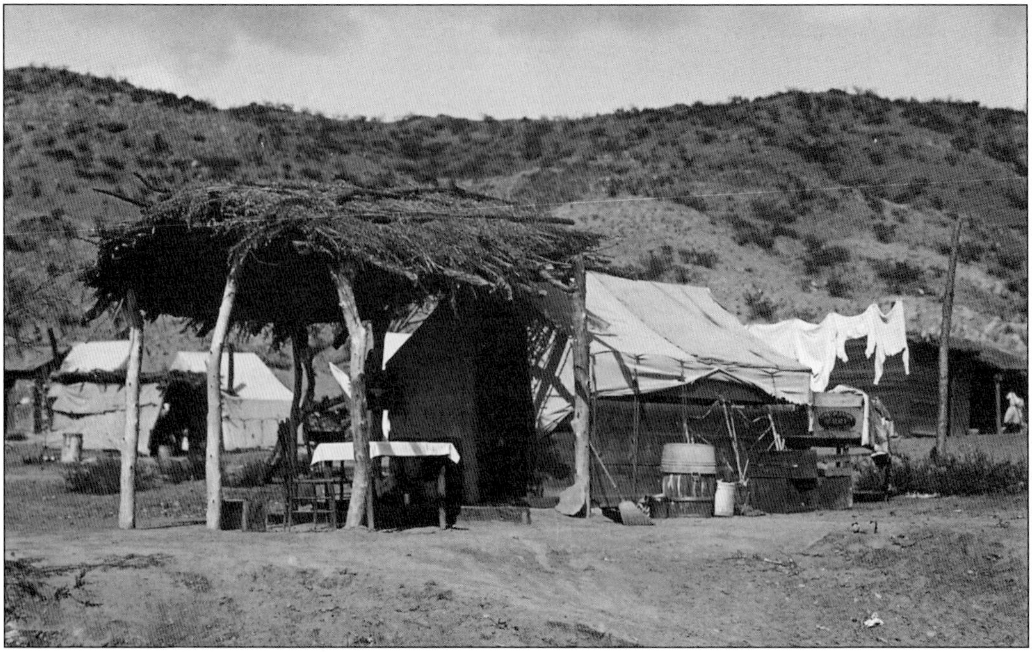

Some of the workers who came to work on the dam, whether they were Americans or Mexicans, brought their families. If they could not afford the government housing, they would construct their own housing of whatever building materials were available. Bunkhouses and a mess hall were provided for the single men, and there were rules within the camp that were strictly enforced. For example, if a man was caught engaging in "excessive drinking," he was dismissed from his employment at the damsite. A sanitation officer went around the camp daily, policing the "wastes" of all kind—human, stable, and kitchen. It was mandatory that the waste be removed daily from the camp. Exercise was encouraged, and an enterprising local resident hauled water from a spring in the Williamsburg area and sold it for 5¢ a bucket.

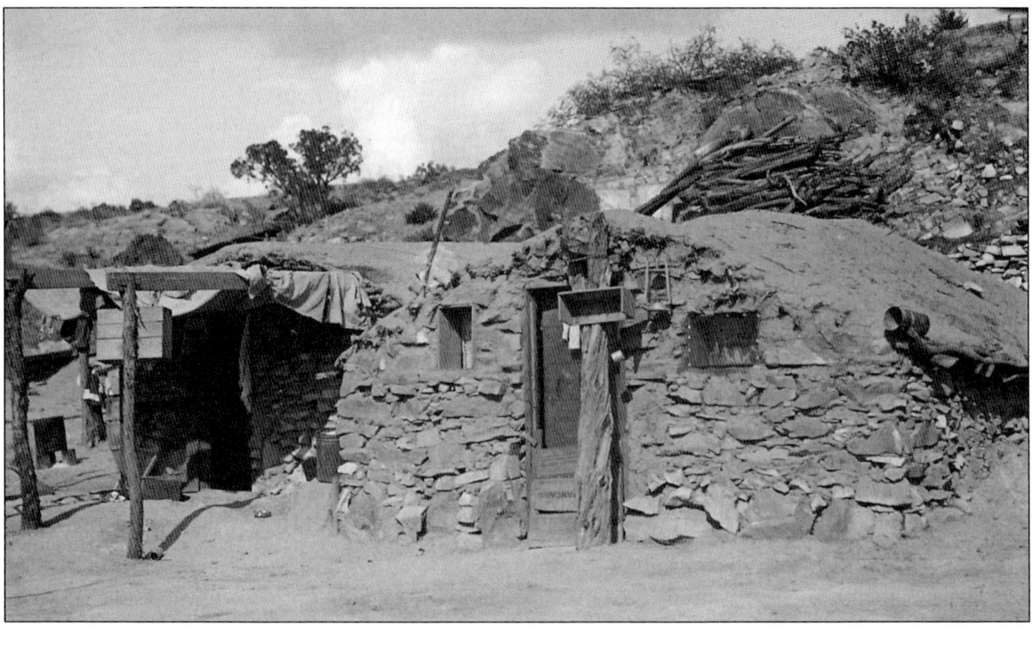

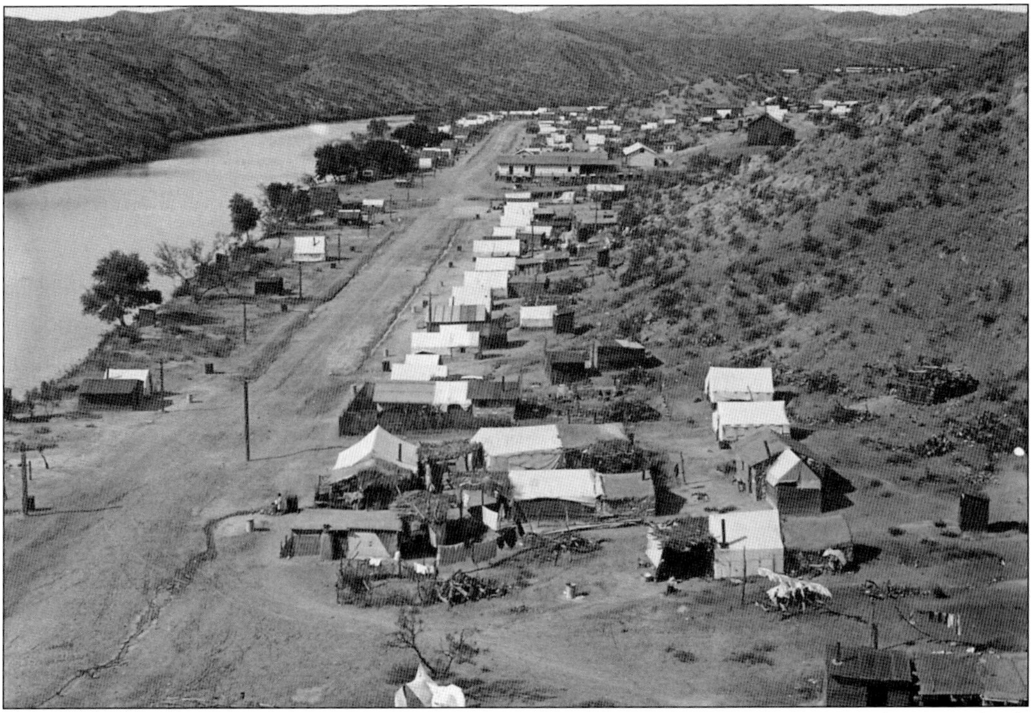

Pictured here are the Lower Town camp and an example of the tent bunkhouses that were built with half wooden sides. A pumping plant had been built in order to supply water to the Lower Town camp from a concrete water tank on the hill. The mess hall was 42 by 126 feet, and it was able to seat 200 individuals, who had to pay 25¢ per meal. Ventilation was provided by fans and louvers. Part of the mess hall had a 65-by-42-feet hall that was originally built to be a dining room but later was converted to the YMCA headquarters. The tents could accommodate from two to eight individuals, and four bunkhouses of similar style were said to have been built for the Mexican laborers, accommodating eight to each tent with double-decked cots.

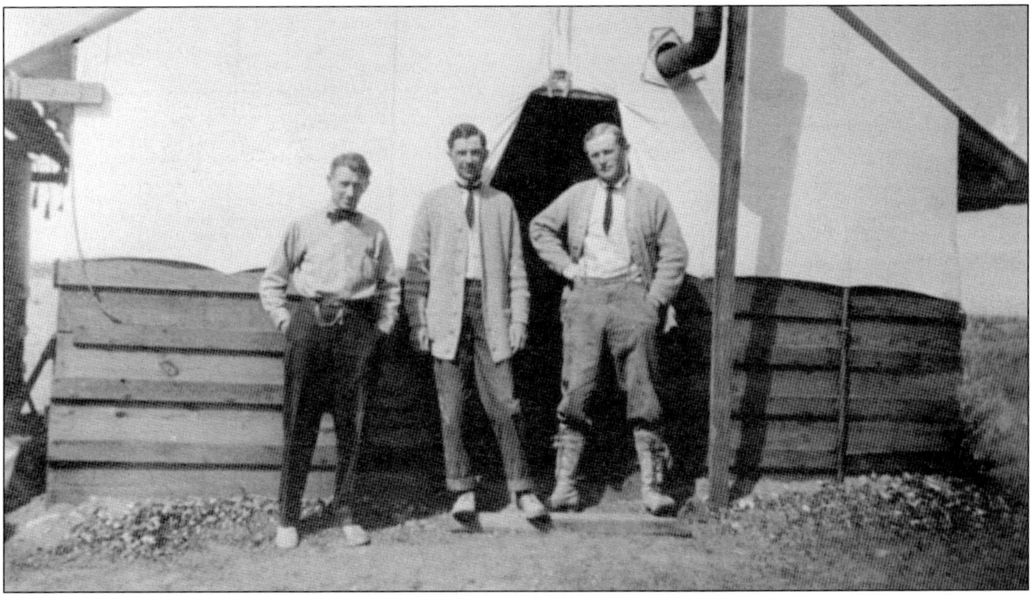

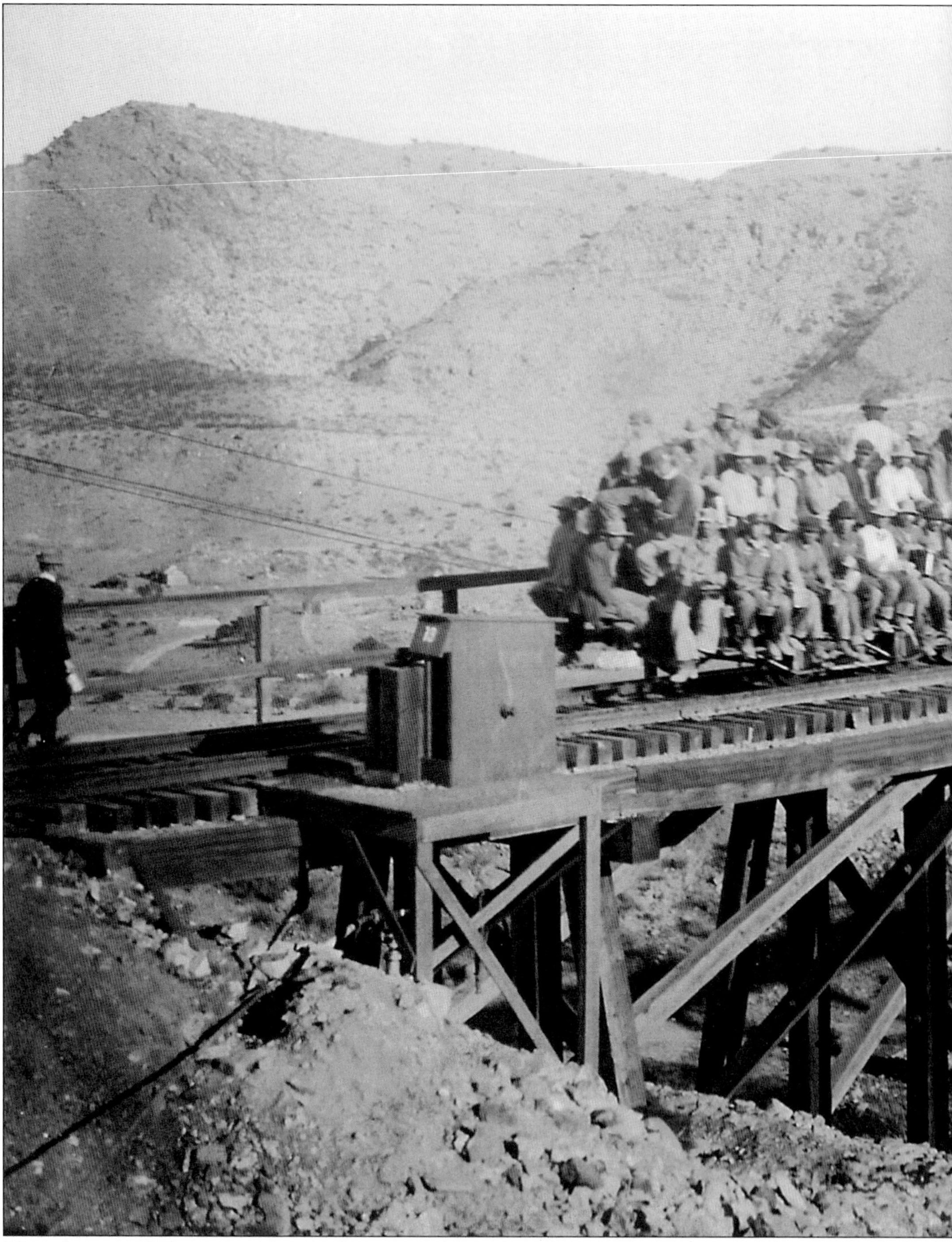

This photograph from January 1914 shows the train called the Elephant Butte and Quarry Special, the "fast mail train to the quarry." At first, the railroad would not allow passengers to ride the train out to see the dam during construction. This changed in due time, and many visitors traveled to

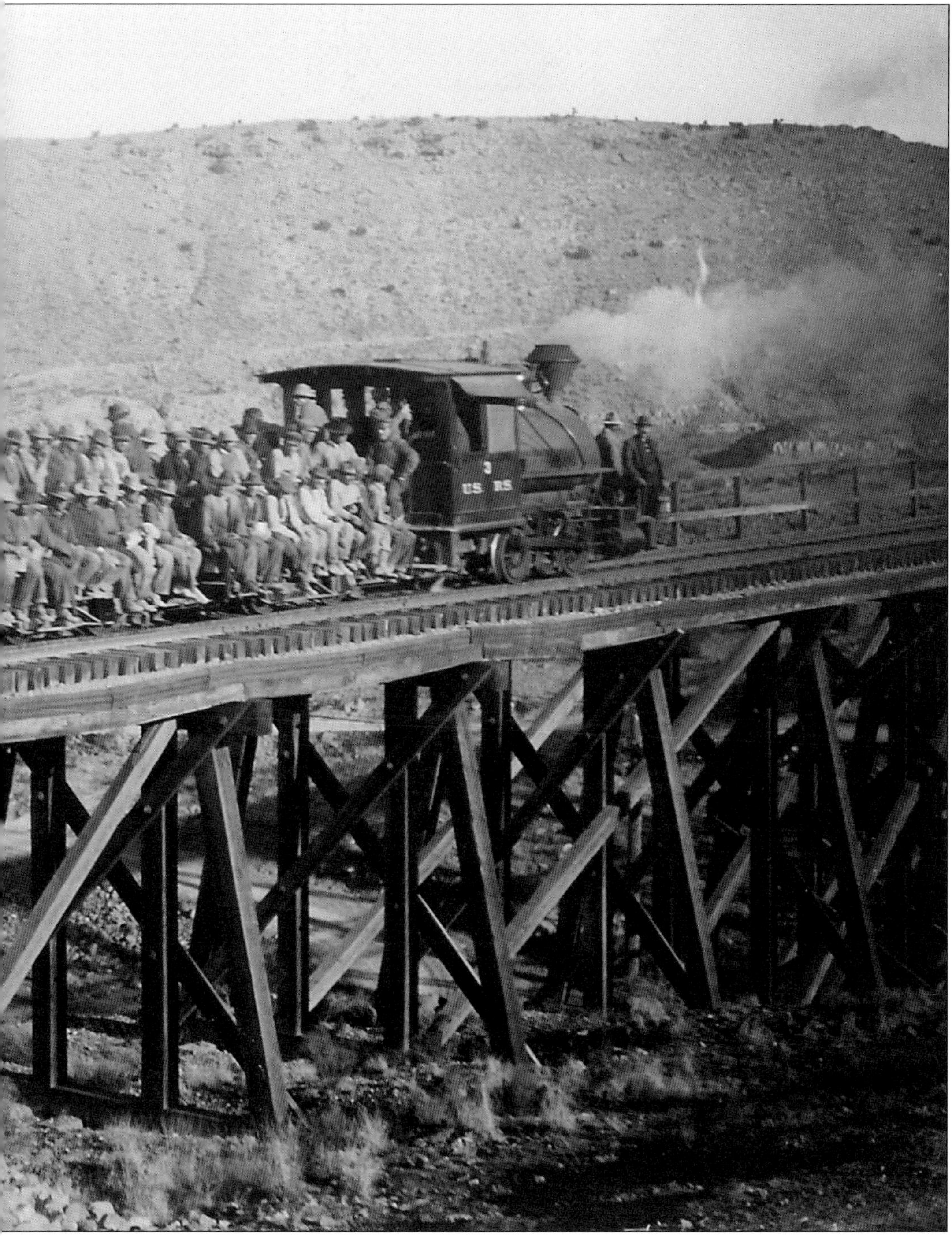

the damsite and surrounding areas by railcar. The train also carried supplies and mail to Elephant Butte damsite three times a week. This train looks as if it might have also carried many of the workers to various parts of the construction.

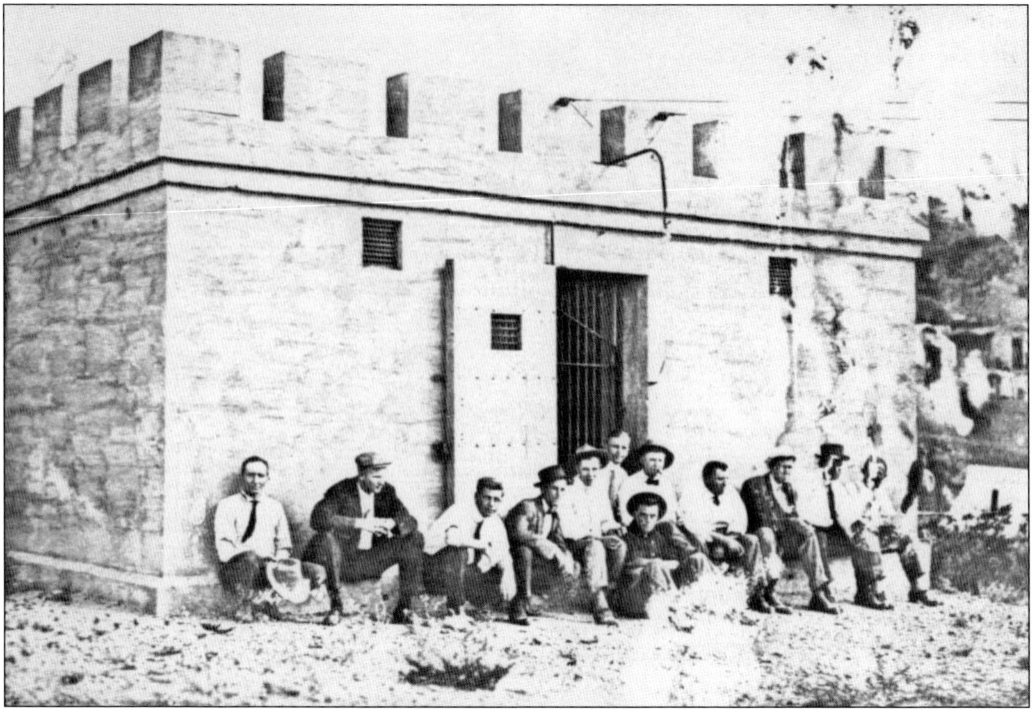

This building was used as a jail. Note the bars on the doors and windows. During the building of the Elephant Butte Fish Hatcheries around 1935, the building would be utilized to store explosives. The individuals in this photograph are unidentified, but they do not appear to be jailbirds.

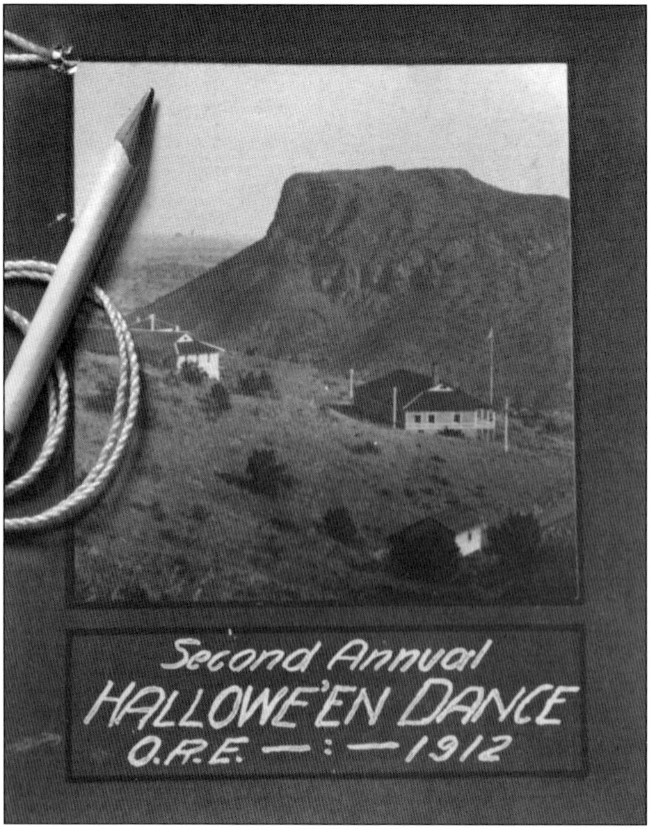

The Halloween dance on October 26, 1912, was called the Order of Royal Elephants. The three patronesses of the evening were Mrs. Blue, Mrs. MacBride, and Mrs. Powell. It is possible that Mrs. Blue was actually Dr. Graham's mother-in-law. Mrs. Blue had lived with Dr. Graham, his wife, and young son in Artesia, California, taking care of the ill Madge Graham. (Courtesy of Moshe Koenick.)

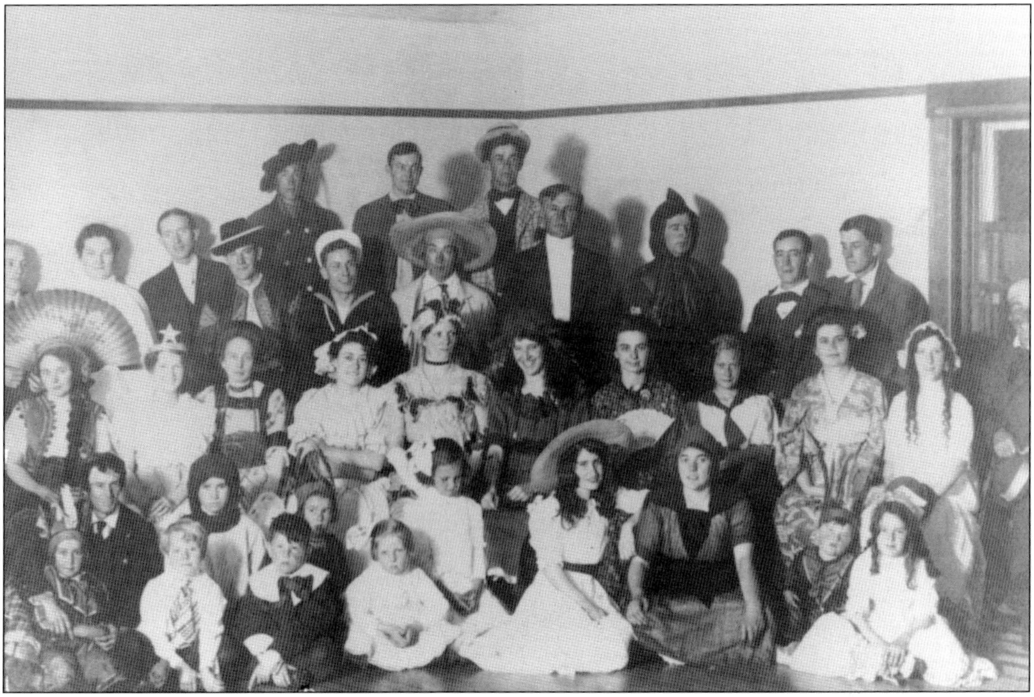

Pictured here is the Flashlight of Masquerade dance given by the Order of Royal Elephants on December 30, 1911. Below is the back of the photograph, which lists the names of the individuals at the dance. As noted by the names and occupations, these residents of what were called the official cottages would hold their ball in the recreation room of one of the buildings on the hill. Individuals from the Lower Town camp also had dances. The so-called Americans had their balls at their hall in a dining room, and Mexican and native New Mexicans of Mexican descent held their celebrations at another location. It was said that there was some rivalry between the Spanish-speaking people of New Mexico and some from Old Mexico. (Both photographs courtesy of Geronimo Springs Museum.)

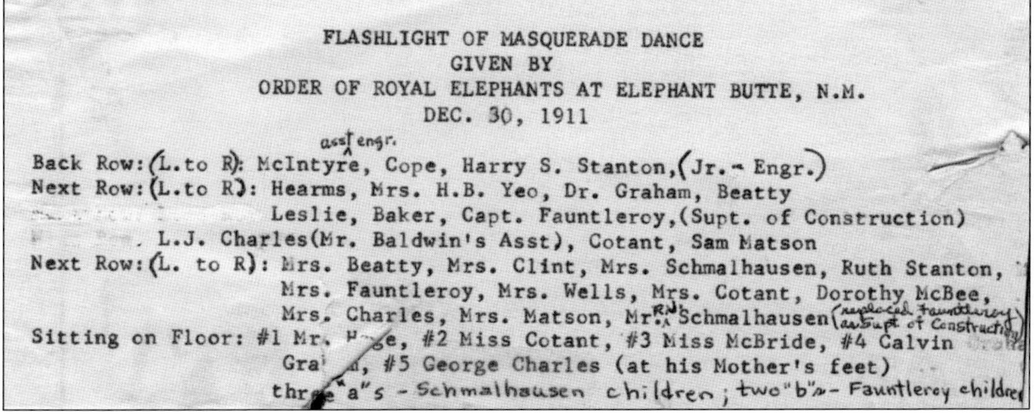

FLASHLIGHT OF MASQUERADE DANCE
GIVEN BY
ORDER OF ROYAL ELEPHANTS AT ELEPHANT BUTTE, N.M.
DEC. 30, 1911

Back Row: (L. to R): McIntyre (asst engr.), Cope, Harry S. Stanton, (Jr. - Engr.)
Next Row: (L. to R): Hearms, Mrs. H.B. Yeo, Dr. Graham, Beatty
Leslie, Baker, Capt. Fauntleroy, (Supt. of Construction)
L.J. Charles (Mr. Baldwin's Asst), Cotant, Sam Matson
Next Row: (L. to R): Mrs. Beatty, Mrs. Clint, Mrs. Schmalhausen, Ruth Stanton,
Mrs. Fauntleroy, Mrs. Wells, Mrs. Cotant, Dorothy McBee,
Mrs. Charles, Mrs. Matson, Mr. R.N. Schmalhausen (replaced Fauntleroy as Supt of Construction)
Sitting on Floor: #1 Mr. H[illegible]e, #2 Miss Cotant, #3 Miss McBride, #4 Calvin
Gra[illegible], #5 George Charles (at his Mother's feet)
three "a"s - Schmalhausen children; two "b"s - Fauntleroy children

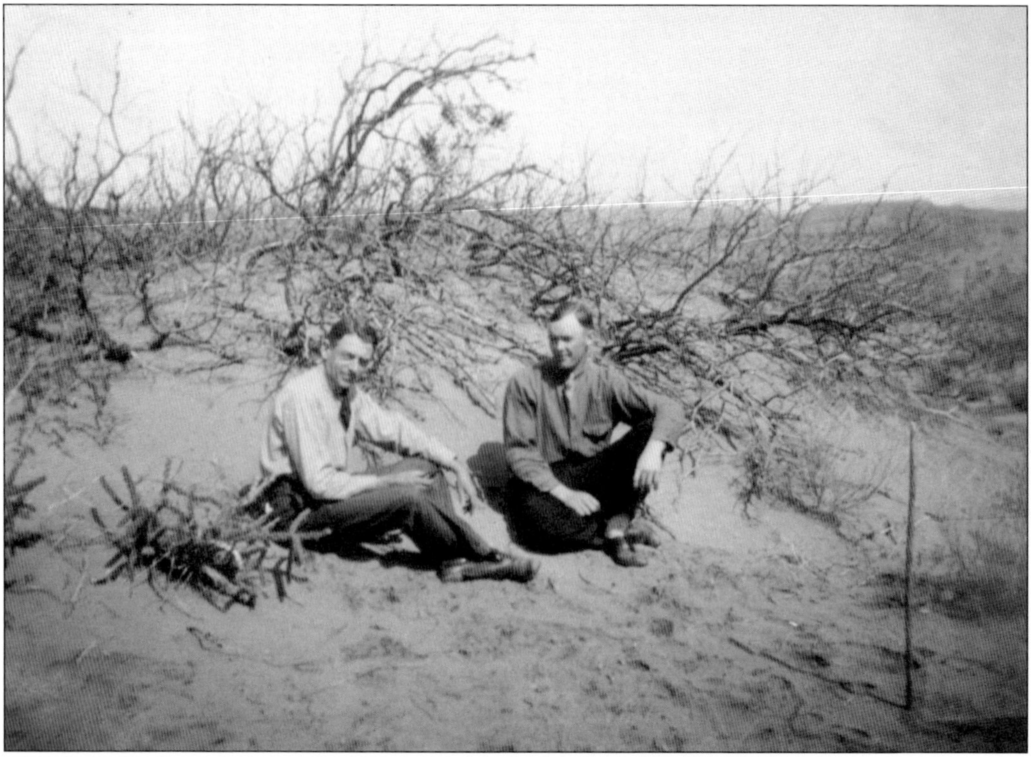

The individuals in each of these photographs are said to be workers on the dam. They are not dressed as workers but appear to be in clothes that would suggest they were visiting the sights. The *Courier-Crescent* of Ohio ran an interesting letter from a visitor to the dam in their issue of November 3, 1916: "We can see Elephant butte and the peak sticks right out of the water like an elephant laying down . . . The dam is honey-combed with passages inside of it, through which we went; it don't look so big looking at it, but when you walk back up the stairs it seems as though you would never reach the top . . . My only regret is that others could not have been with me to have enjoyed themselves as I did." (Both photographs courtesy of Geronimo Springs Museum.)

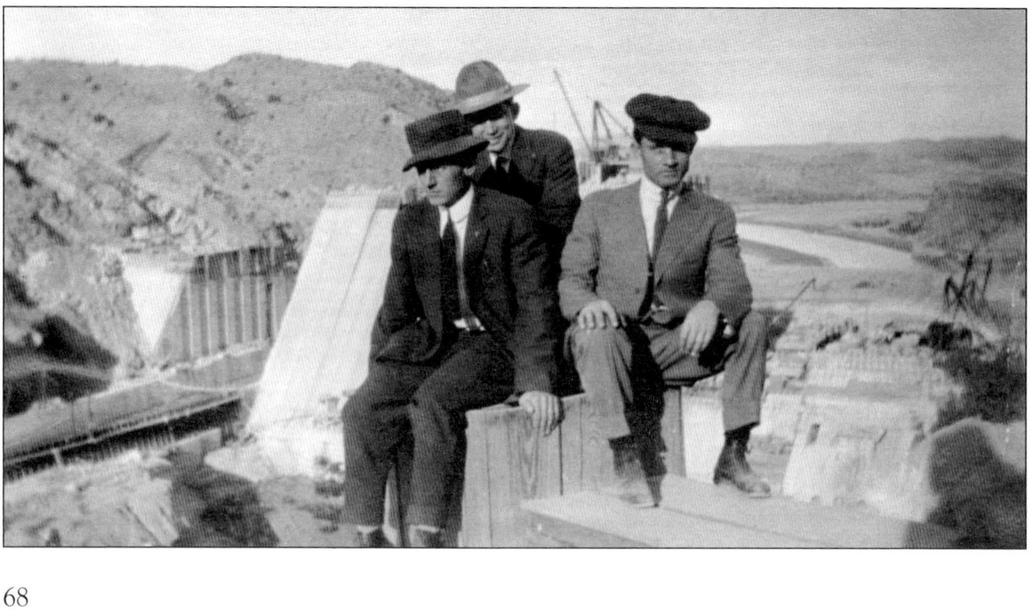

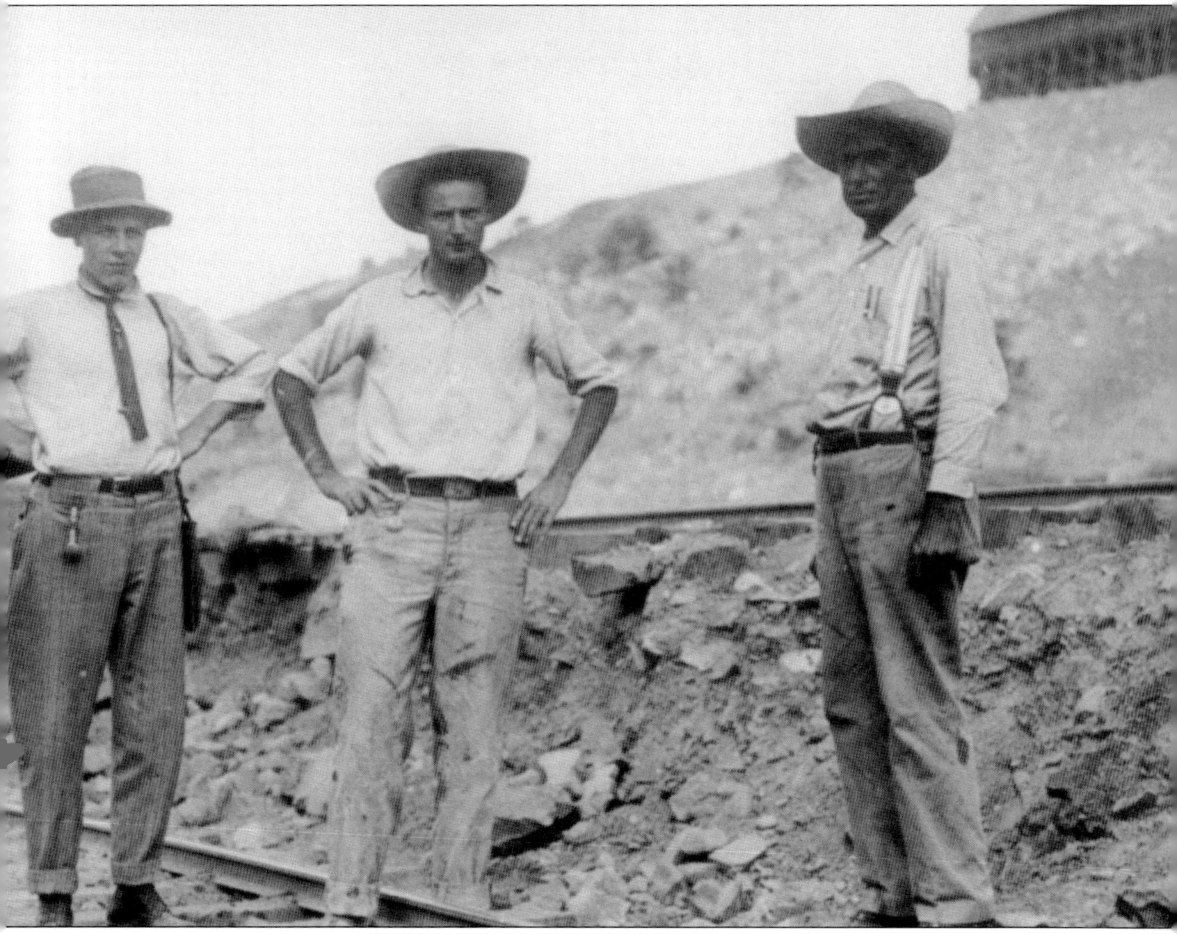

Robley J. Schmalhausen (far right) held the position of superintendent of construction. He lived with his family in the Upper Town until the dam was built then moved to El Paso, Texas. In 1924, he returned to Elephant Butte to act as reservoir superintendent. Schmalhausen died in 1932 in the vicinity of Elephant Butte Dam when he lost control of his car and plunged to his death down a steep arroyo. A week after Schmalhausen's death, Edward Romage, a former employee of Schmalhausen, went to the accident scene. The *Manitowoc Herald Times* of October 14, 1932, noted that "Friends said he [Romage] looked down into an arroyo into which Schmalhausen's car plunged" and lost control of his car, rolling 300 feet down the same arroyo that had taken the life of Schmalhausen. Interestingly, the paper noted, "A desire to see the spot where a fatal automobile accident occurred causing the death today of Edward Rommage, 65, at the same place and in the same manner."

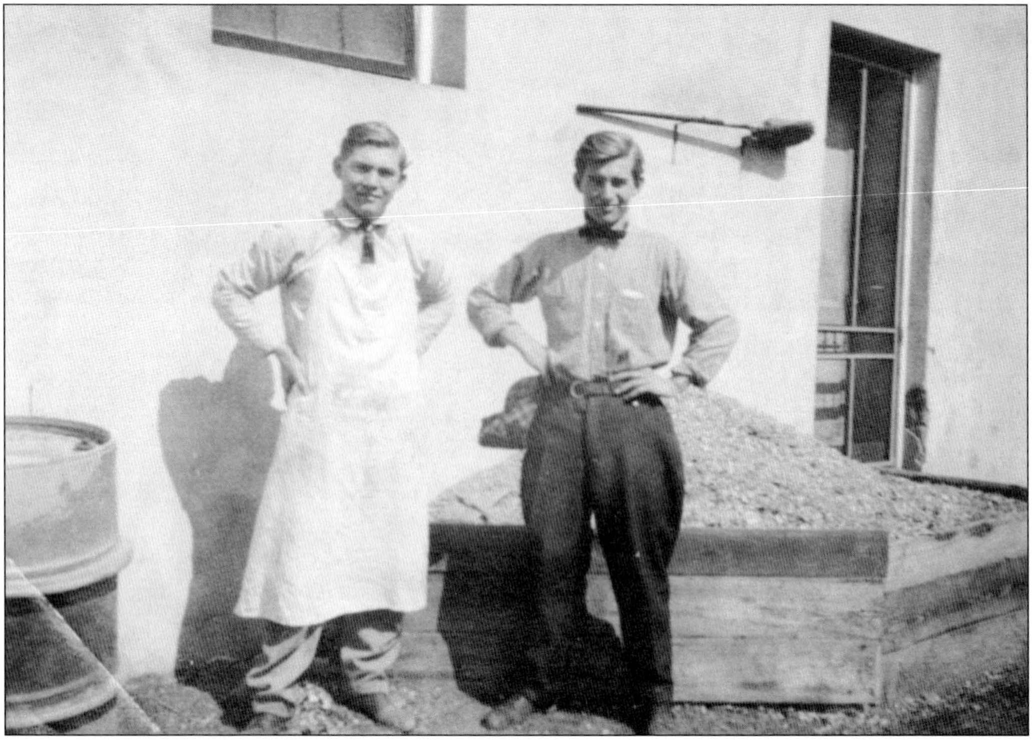

The commissary, 57 by 77 feet in diameter, was under government management and operated as the only store in camp. Prices were said to be reasonable, with fresh fruits and vegetables kept in stock. There was a meat market that was under "rigid inspections," according to Dr. Graham. The icehouse and cold storage were housed under the same roof, and the commissary employed 10 men. As pictured below, the lower camp had houses that were made of canvas, wood, adobe, or a combination of all three. Along with the tent-style bunkhouses, there were three houses that accommodated 64 men in double-decked cots. The bunkhouses had screens for ventilation and electric lights, water, and shower baths. Three bathhouses, 12 by 16 feet each, were equipped with hot water, two showers, and one tub. Only the so-called "Americans" were able to use the bathhouses.

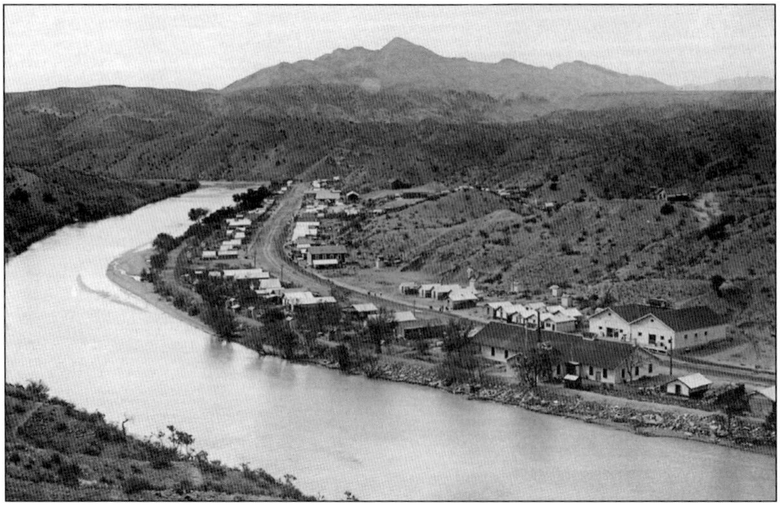

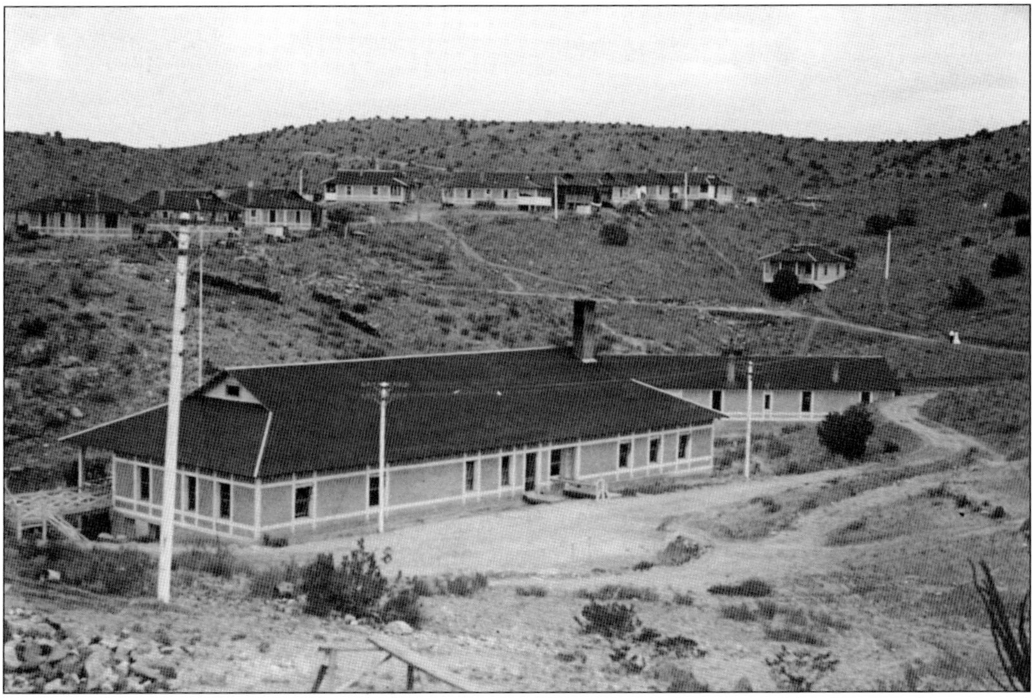

The part of camp called the Upper Town was where the engineers and office workers lived. The Upper Town contained the main office (shown above), a quarterhouse, mess hall, chemical laboratory, and cottages. The quarterhouse, seen below, was 34 by 145 feet and stood two stories high. It could accommodate about 50 people. The quarterhouse had steam heat, electric lights, bath and sewer connection, and a large dancing floor. It also had card, reading, and billiard rooms for the entertainment of the boarders. All the buildings in the Upper Town were said to be pebble-dashed adobe construction. There were 14 cottages in the Upper Town that had three, four, and five rooms, and all of the cottages were equipped with electric lights and bath and sewer connections.

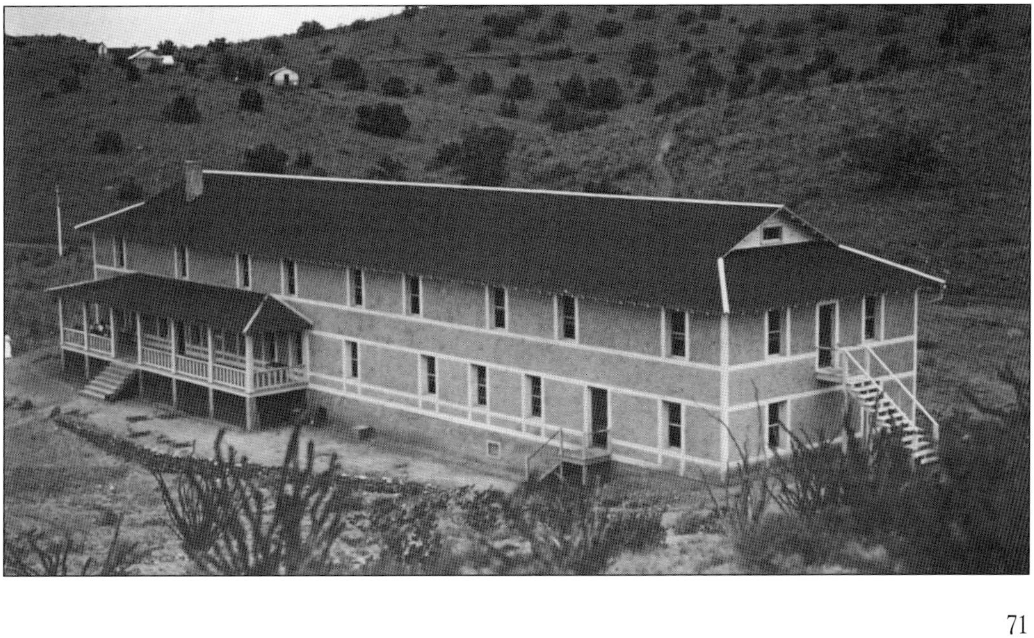

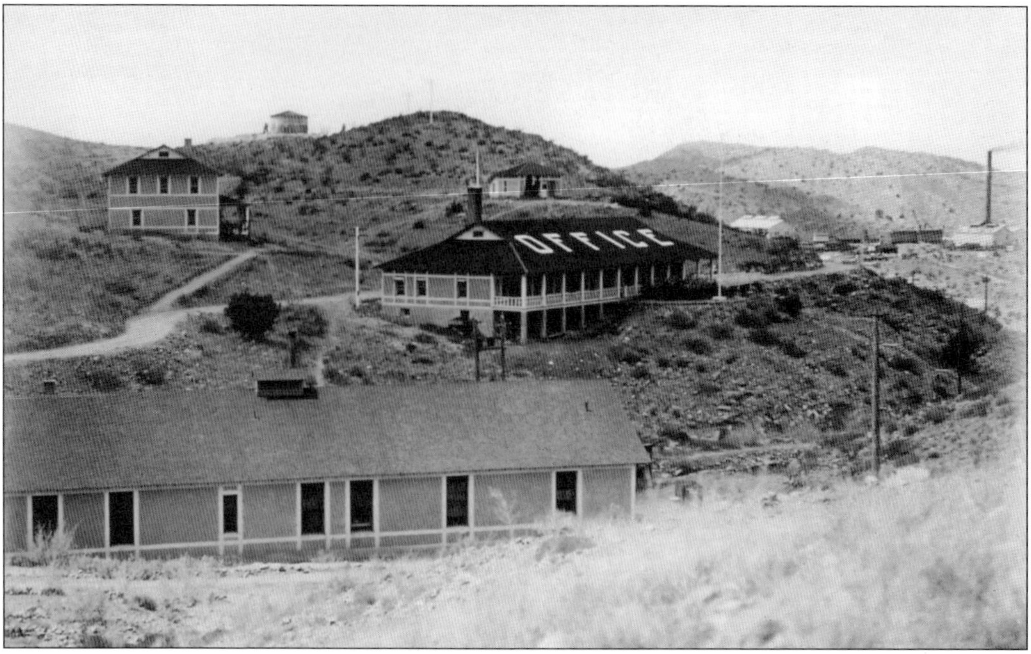

The office was the official administration building for Elephant Butte Dam. It was 100 feet by 49 feet and included a 10-foot porch. It was built on a slope with the main floor being at grade on the building's west side. The foundation was concrete, and an interior brick chimney protruded from one end of the roof. The main room had a drafting room, printing room, vault, telephone room, storeroom, chief clerk's office, disbursing room, timekeeper's room, and three engineer's rooms. The southern basement housed the furnace, which vented through the chimney. The only access to the building was through a single door on the west side of the building.

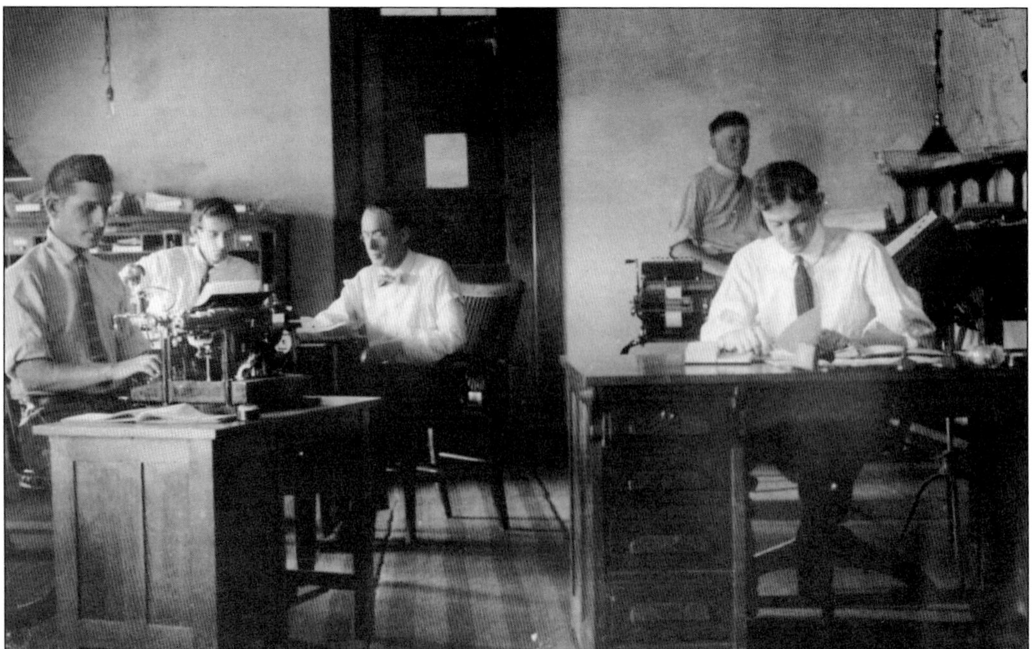

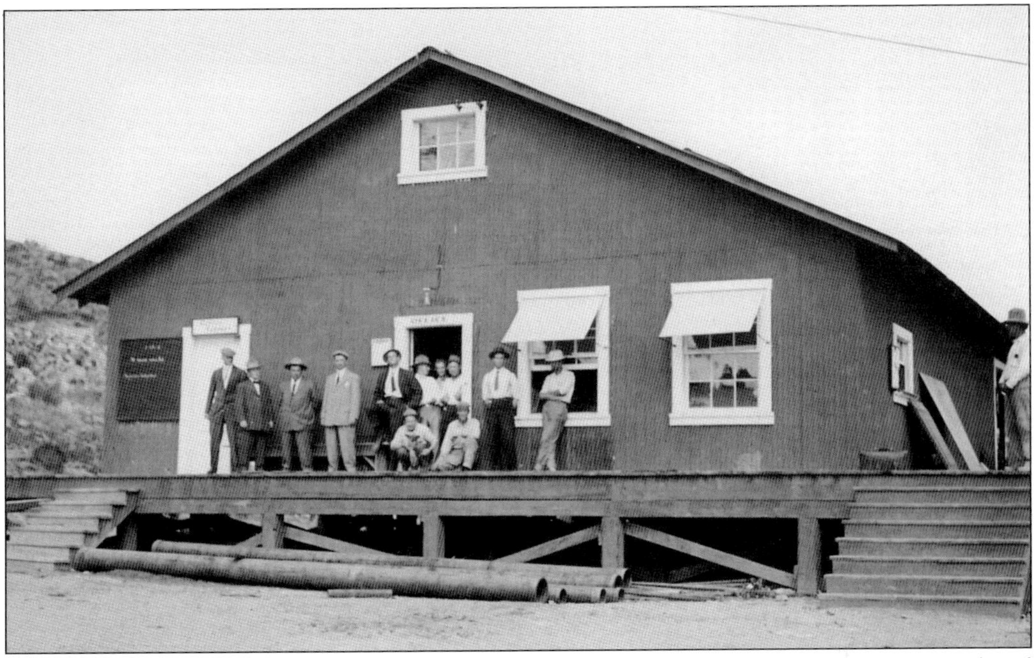

The photograph above was labeled "warehouse #4," and the image below features the railroad station at the townsite. The townsite was a small, self-contained village, with telephone poles carrying electricity down both sides of the street. There was a big issue with flies at the Lower Camp, blamed mostly on the horse corrals. Fly traps were placed everywhere in the camp, and it was claimed that millions of flies had been caught. The Upper Camp was not plagued by the flies, as there were no corrals located in the Upper Camp. Dr. Graham believed that the ceiling of the fly traps should be painted white and the floors a darker color so the fly would move to the point of "greatest light."

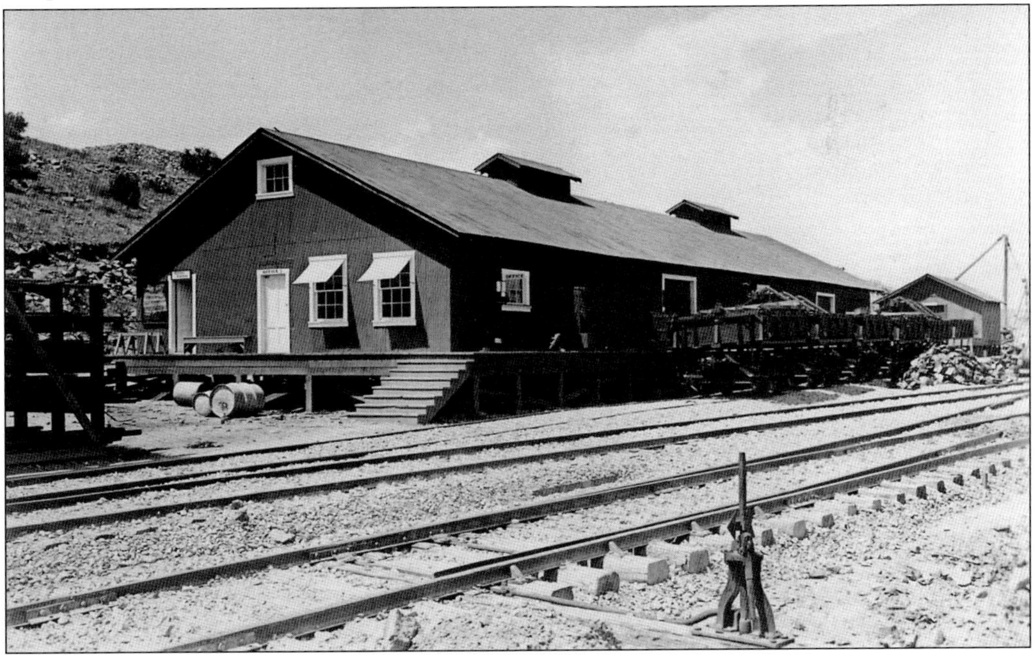

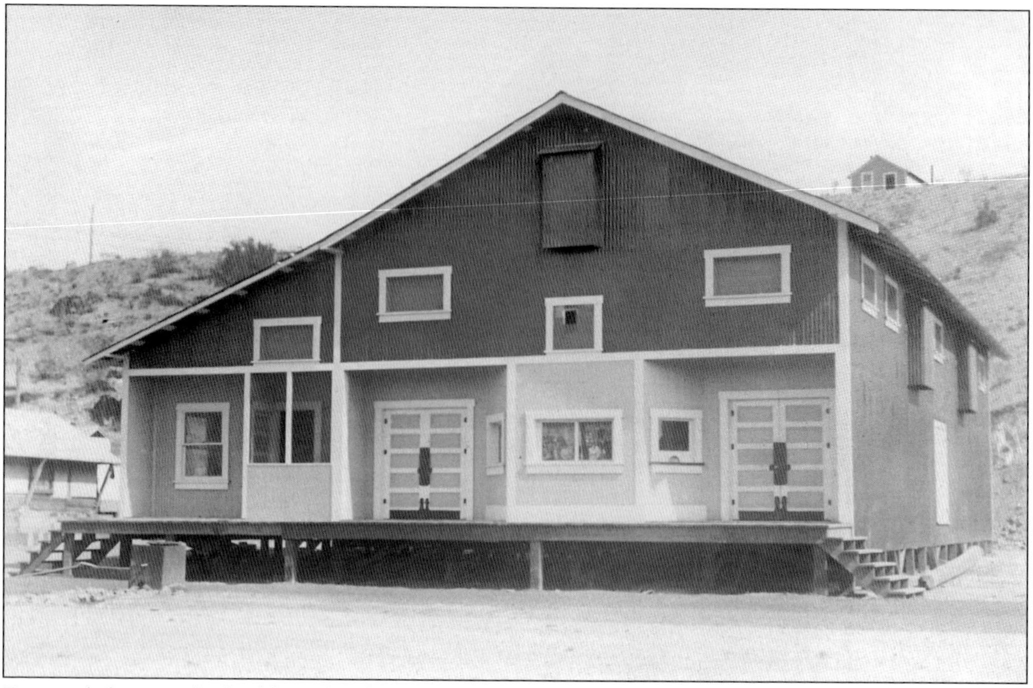

Pictured above is the building that housed the moving-picture theater and ice-cream parlor. It had been built, according to Dr. Dale Graham, solely for the entertainment of the camp occupants. It was operated by the US Reclamation Service, and Dr. Graham claimed that only sufficient revenue to meet expenses was expected and needed. Pictured below are the tennis courts, located in the Upper Town. It was said that a "high-class amateur orchestra" gave frequent free concerts, baseball teams were organized and enjoyed by all, and tennis and croquet grounds were well maintained. Children in the Upper Town had the luxury of being involved in Boy Scouts, and Dr. Graham further noted that "Secret Orders have built lodge halls."

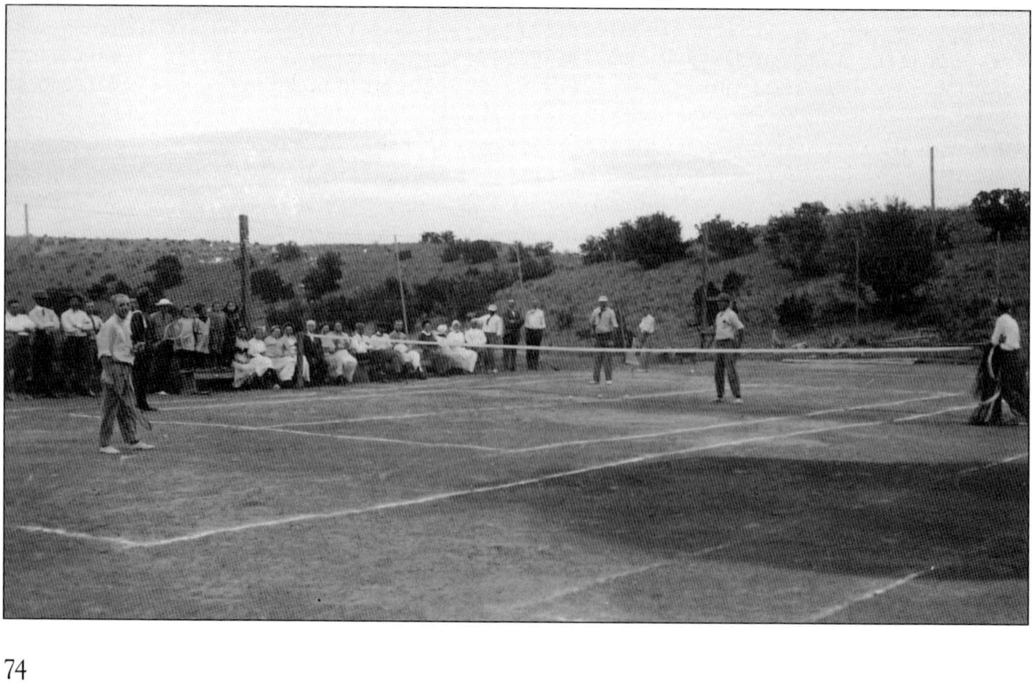

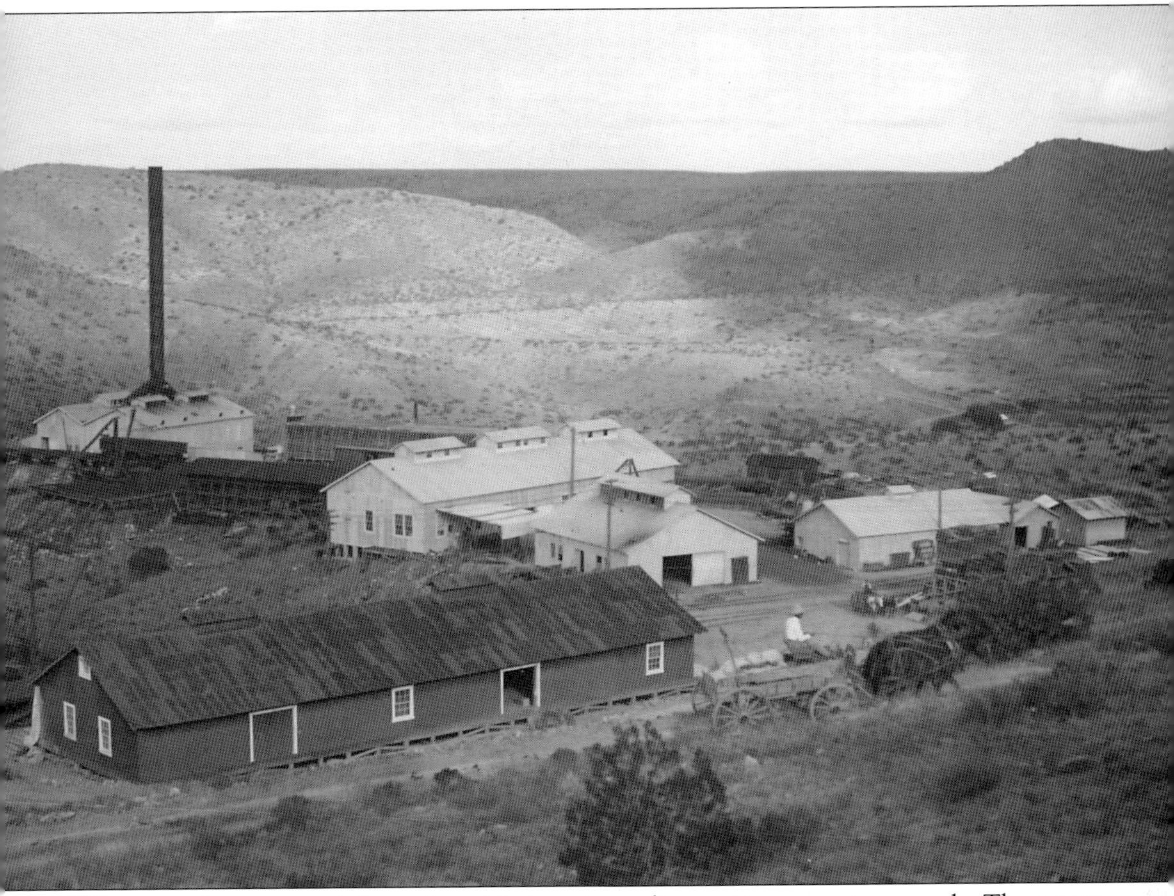

Whenever there was an injury of any kind, an immediate written report was made. The report included the name, age, kind of work, salary, character of injury, any witnesses or circumstances surrounding the injury, and other pertinent information. The report was made in duplicate, and thorough records were maintained. Dr. Graham was required to make not only a 10-day report but also monthly and yearly reports that covered the health of the camp. Every employee had a mandatory deduction on each paycheck of $1 per month. This deduction entitled the employee to what Dr. Graham referred to as "all reasonable dispensary or hospital service." He claimed that there was practically no abuse of this "privilege," and the hospital funds were actually in the black. All employees had to agree to an examination of physical fitness. Despite contractors saying this would handicap their efforts to secure labor, Dr. Graham noted that only 3 out of over 3,000 applicants had refused to be examined as part of the job requirement.

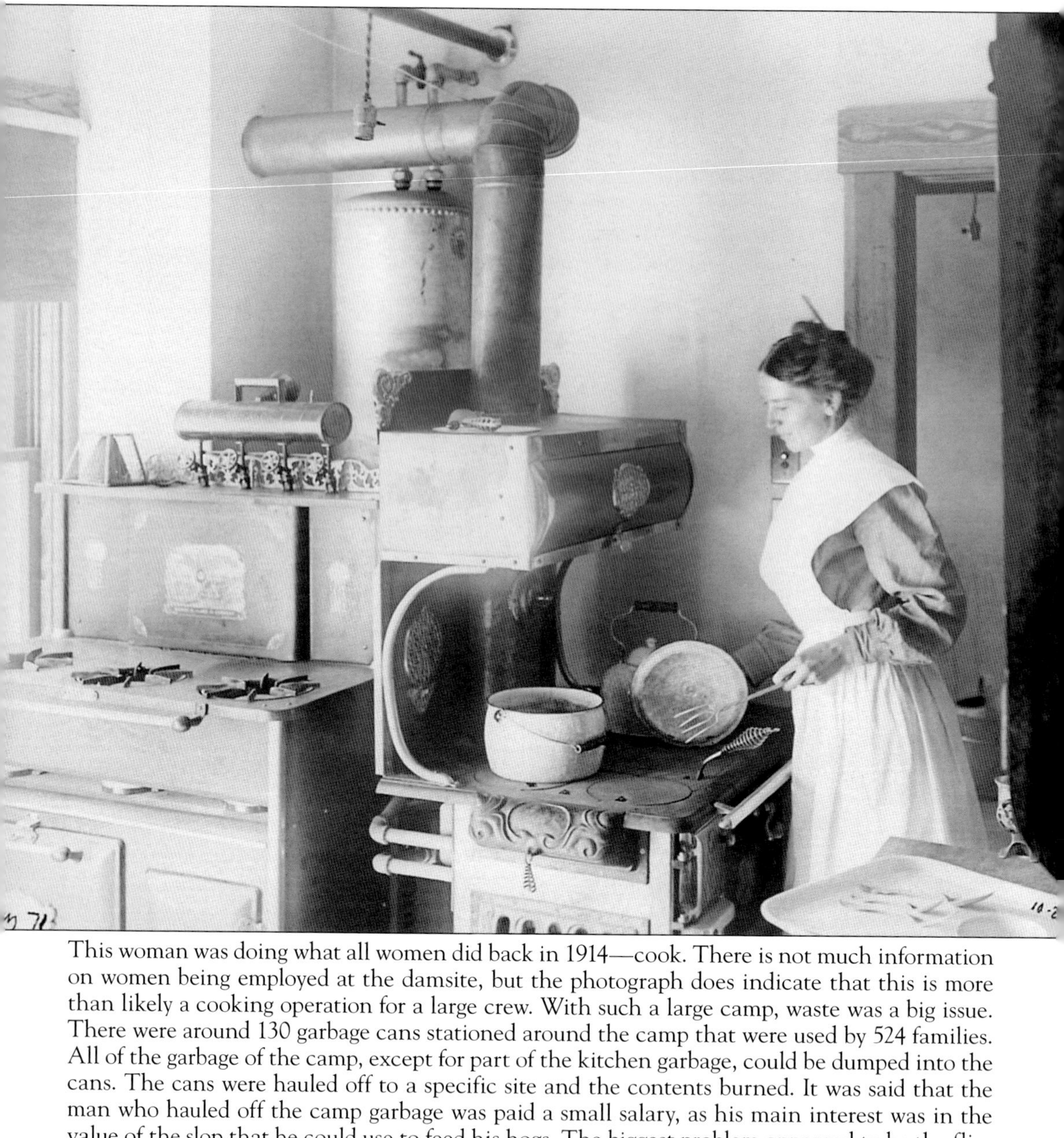

This woman was doing what all women did back in 1914—cook. There is not much information on women being employed at the damsite, but the photograph does indicate that this is more than likely a cooking operation for a large crew. With such a large camp, waste was a big issue. There were around 130 garbage cans stationed around the camp that were used by 524 families. All of the garbage of the camp, except for part of the kitchen garbage, could be dumped into the cans. The cans were hauled off to a specific site and the contents burned. It was said that the man who hauled off the camp garbage was paid a small salary, as his main interest was in the value of the slop that he could use to feed his hogs. The biggest problem appeared to be the flies that the garbage attracted.

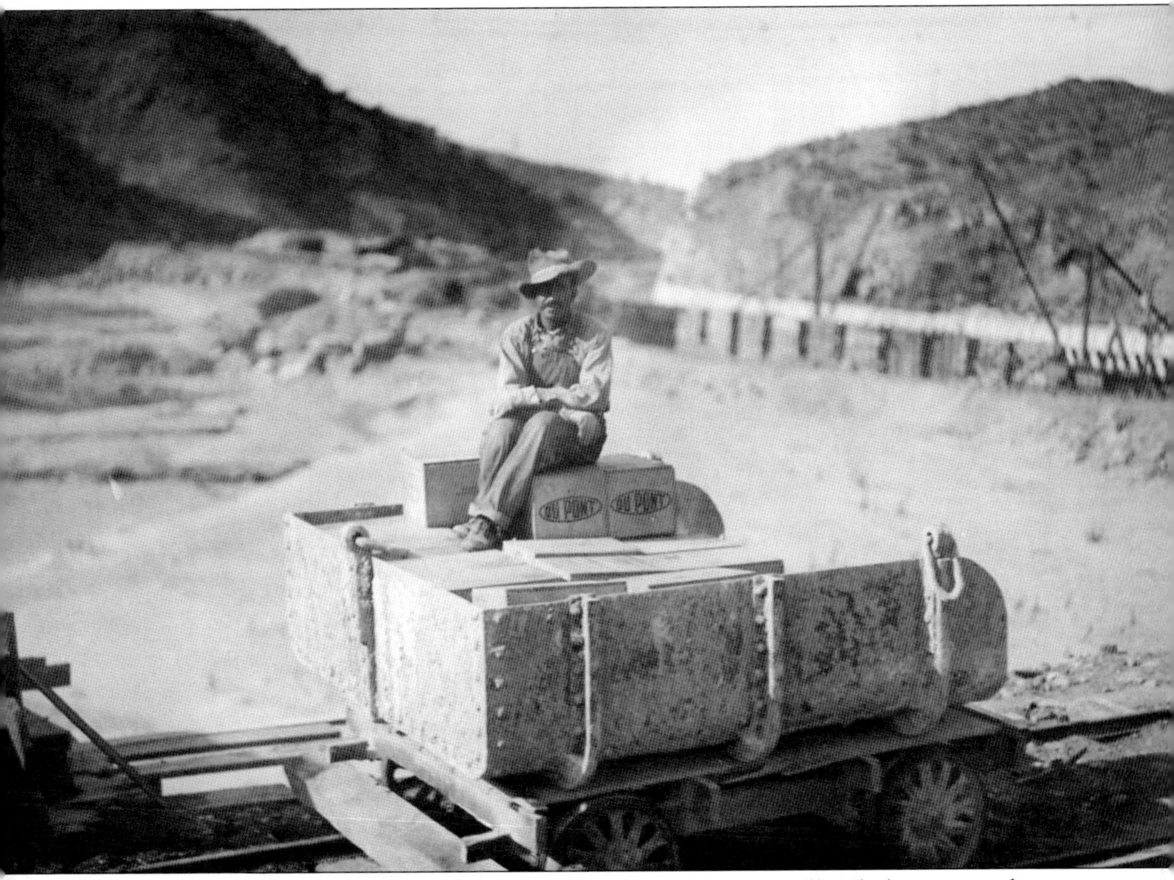

Controversy brewed over whether or not the government was using illegal aliens as workers on the dam, and the issue was investigated by Frederick H. Newell, director of the Reclamation Service. Newell noted that any person "claiming American citizenship" was welcome to work and emphasized that it did not matter if he spoke English or not. He said there had not been discrimination, in that the government had used local labor as much as possible. On March 9, 1916, a band led by Pancho Villa crossed the border and killed 16 American citizens in Columbus, New Mexico. On March 18, acting chief of construction L.J. Charles sent a letter to construction engineer E.H. Baldwin saying that he detected a ripple of tension in camp resulting from the incident. He wrote, "Conditions on the border are beginning to be felt in the labor here. There is no danger of trouble unless some drunken specimen of either race should start it; then it is difficult to tell just where it would end."

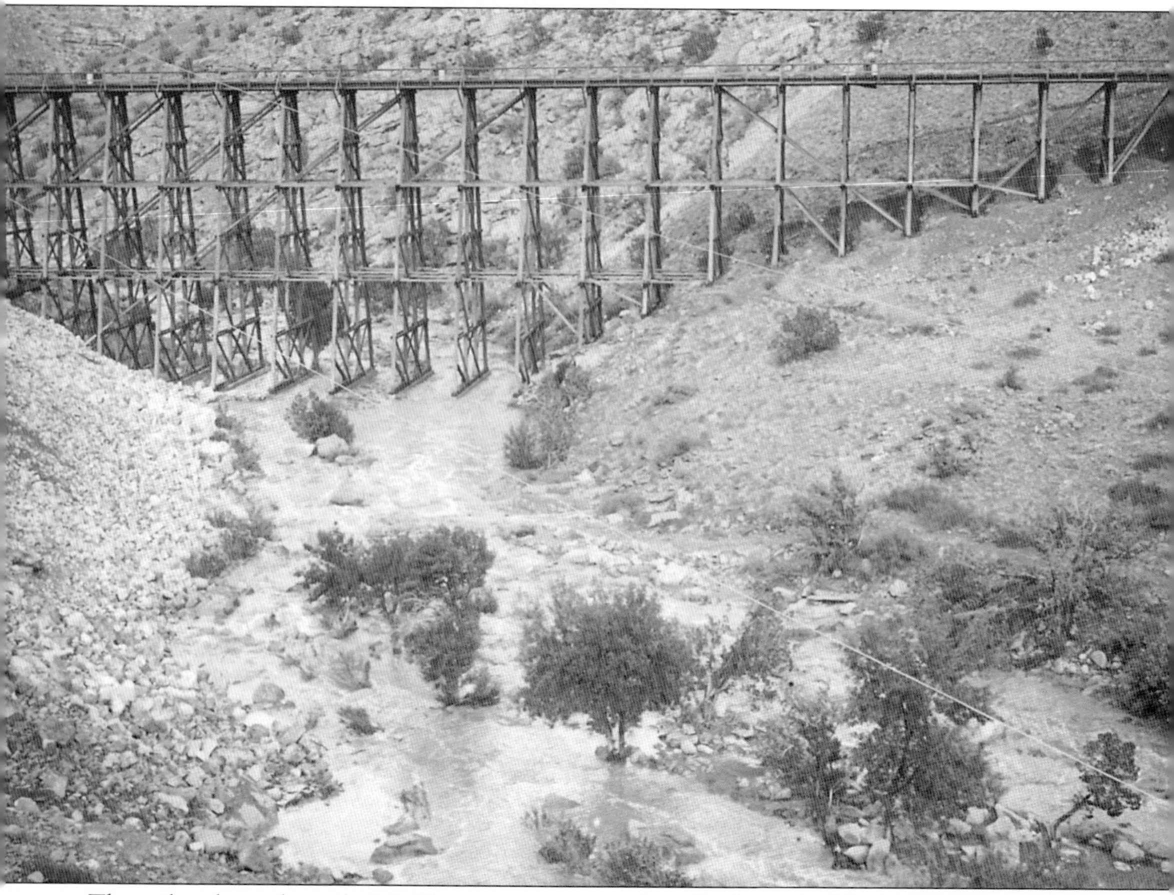

The railroad ran through Ash Canyon on its route from Engle to the Elephant Butte Dam. The bridge, as seen above, was a large wood trestle bridge that crossed the canyon. The railroad spur was built between 1908 and 1911. Ash Canyon was close to the portion of the Lower Camp, which was designated for the Mexican laborers. There was one power line that was extended into that portion of the Lower Camp, enabling electricity. Skilled laborers made from $2 to $2.50 an hour, and workers of all nationalities could make $1.25 to $1.50 a day for those unskilled jobs in concrete mixing, placing, and finishing. Jobs such as blacksmith, plow driver, teamster, and carpenter garnered from $1.25 to $4 per day. Up to 1,200 men were employed at one time at the damsite from 1914 to 1916. It was said that the population of the two camps fluctuated between 2,500 to 3,000 individuals during various times of construction.

Four

Building a Modern Marvel

The Dam

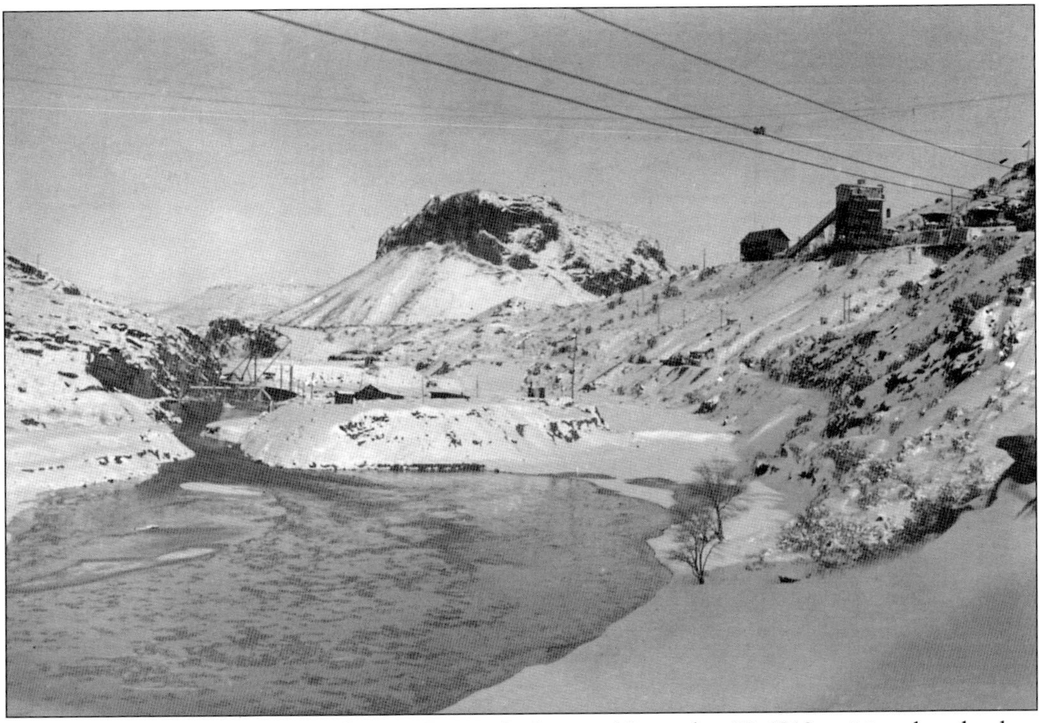

To put it in perspective, the *Washington Post* said it best on November 23, 1913, writing that the dam would "protect the lower valley from destructive floods and at the same time insure an abundant water supply to 180,000 acres of land in New Mexico, Texas and Old Mexico, restoring to cultivation and intensive agriculture thousands of acres abandoned by reason of water shortage."

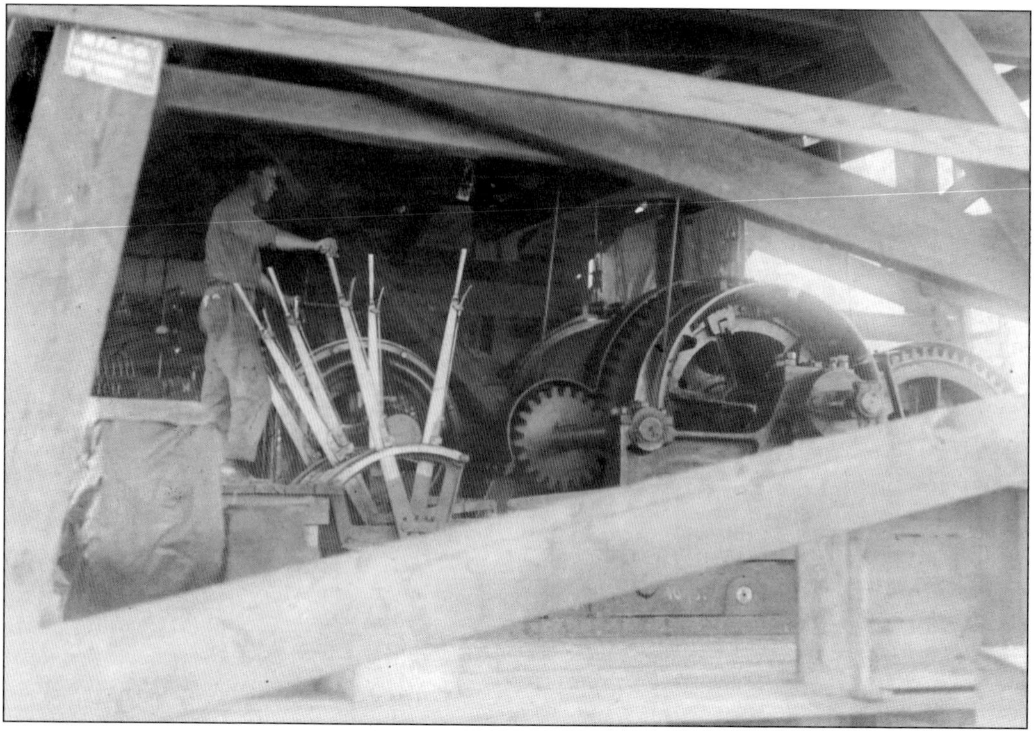

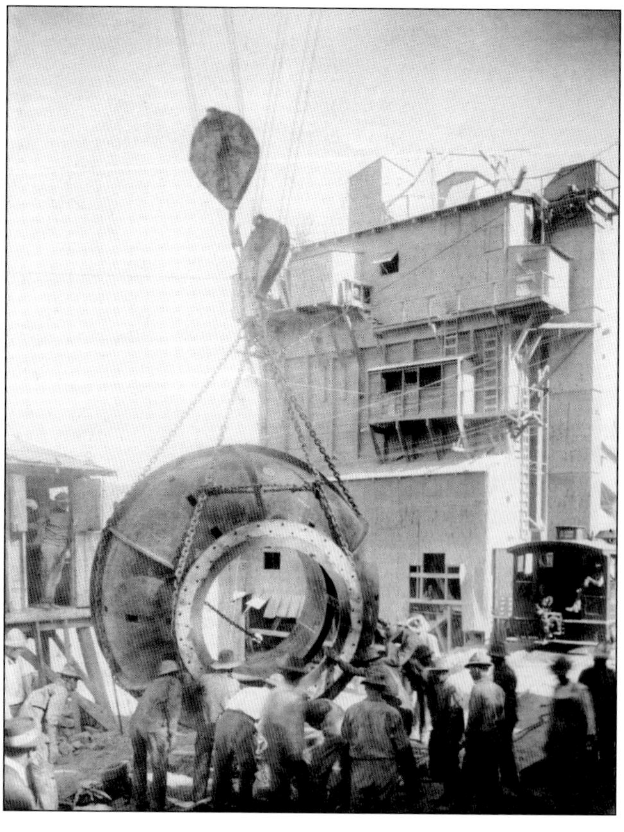

Labor relations at the dam project were not always considered "favorable." There were also times during the construction of the dam that labor was in short supply. Machinery could not always do the hand work necessary for the extensive excavation. The work was not only dangerous but also physically draining. Machinery brought in for the construction of the dam often had to be reassembled. To keep the machinery in good working condition was a full-time job. In 1913, sealed bids were being accepted by the US Reclamation Office in Washington, DC, to furnish the dam project with four 60-inch balanced valves and accessories and four 47-inch-by-90-inch service gates and accessories.

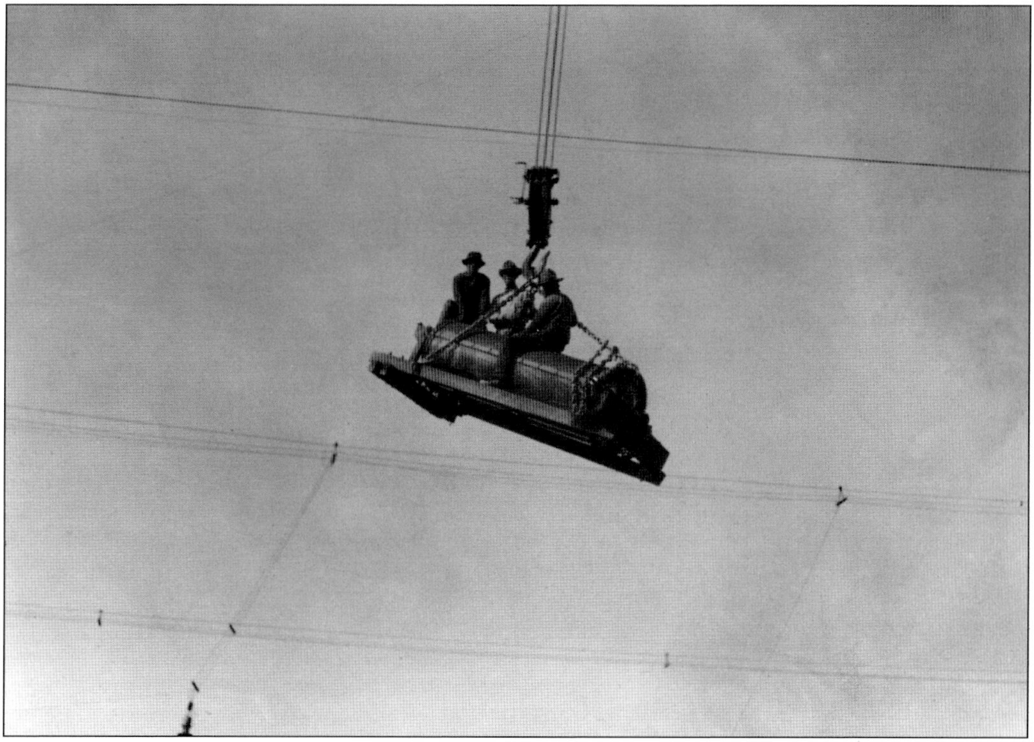

It would appear that the daredevils pictured above could not resist taking the ride of their life. The cableways systems were used extensively to carry heavy loads across the damsite, and differing accounts of how much actual weight the cableways could carry was debated. When the need to take a locomotive (weighing 16 tons) across the river came up, the debate heated up. The locomotive could be brought up using the bridge a mile upstream but the cost in time, money, and labor to construct the temporary track had to be considered. The *Waterloo Times-Tribune* reported on February 26, 1913, "The daring engineers swung it across the river on two overhead cables 1,200 feet long, although either of the cables alone would have snapped under the strain."

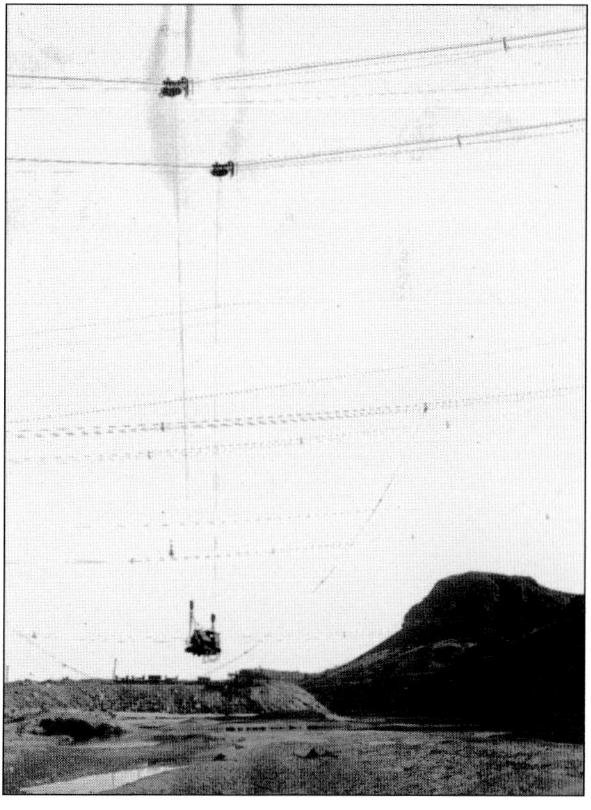

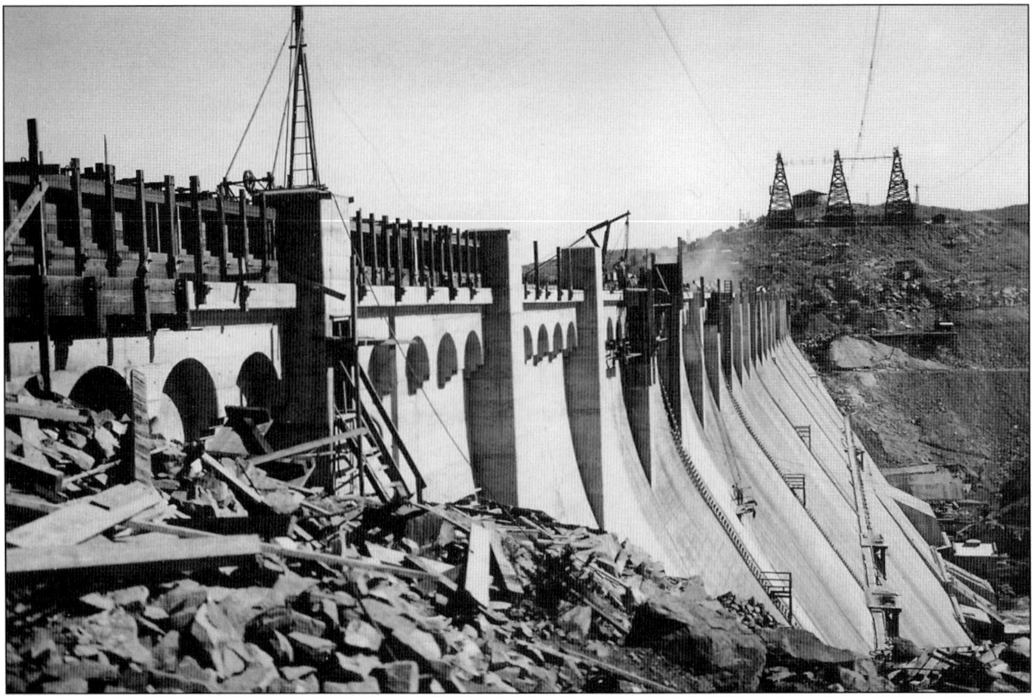

On January 8, 1914, the *El Paso Herald Post* reported that Vicente Cortes Herrera, the undersecretary of communications and public works in Mexico, visited Elephant Butte Dam. He was "expected to look over the river alignment work along the Rio Grande." A treaty with Mexico and the United States had been signed on May 21, 1906, and Article I of that contract said, "After the completion of the proposed storage dam near Engle, New Mexico, and the distributing system auxiliary thereto, and as soon as water shall be available in said system for the purpose, the United States shall deliver to Mexico a total of 60,000 acre-feet of water annually in the bed of the Rio Grande at the point where the head works of the Acequia Madre, known as the Old Mexican Canal, now exists above the city of Juárez, Mexico."

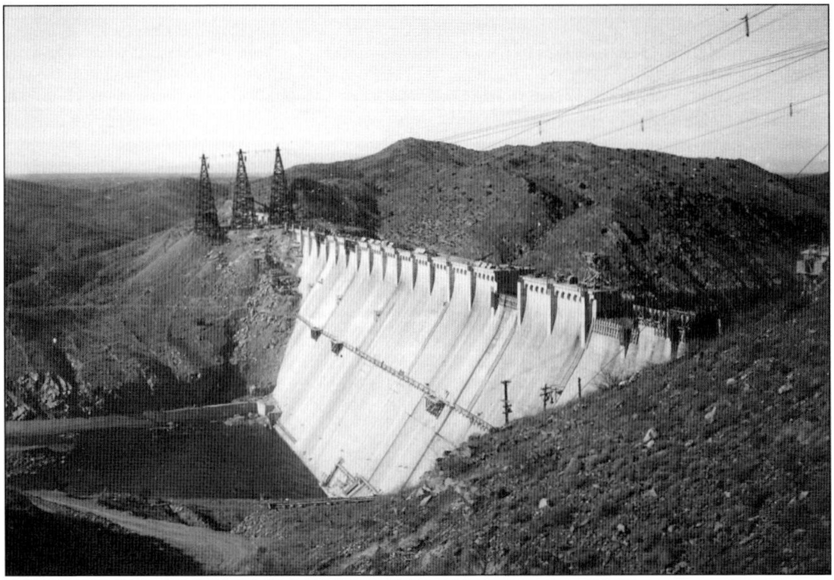

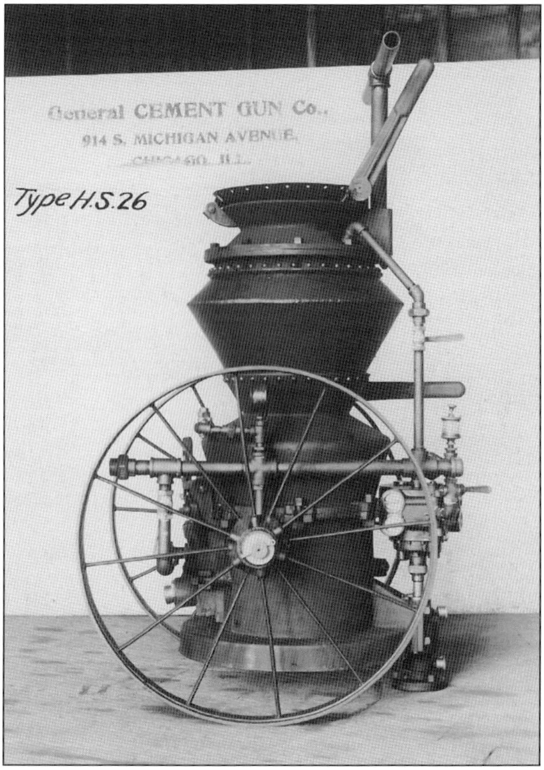

The dam was to be faced with one inch of grouting blown under heavy pressure from a cement gun, making it almost impervious to water. Nevertheless, drainage holes 8, 10, and 12 inches in diameter and eight feet from center to center had been left in two lines the full length of the dam, opening into chambers, which communicate with a large drain-way. An operator in a small wooden raft waterproofed the dam by spraying a mixture of cement, sand, and water on the upstream dam with a gun. The gun, operated by compressed air, coated horizontal strips 10 feet high by the length of the dam to a quarter of an inch thick. Four coats were applied, and the engineers noted that there was perfect adhesion of the waterproofing to the original concrete.

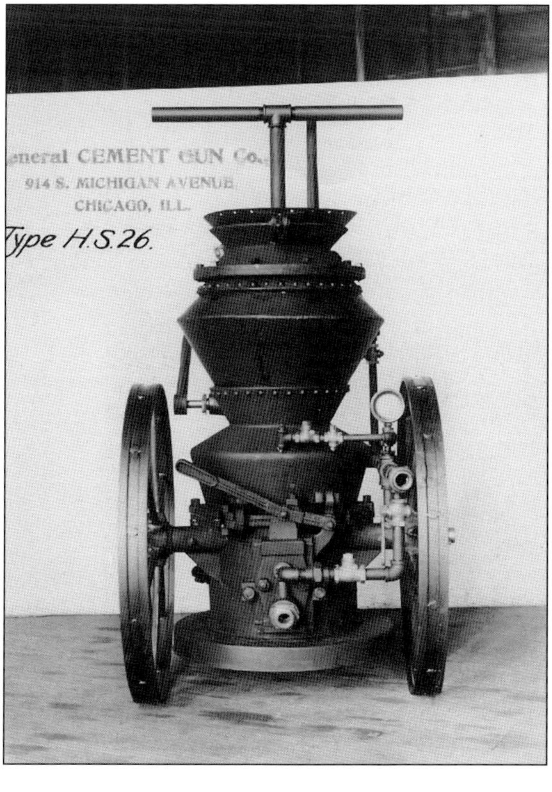

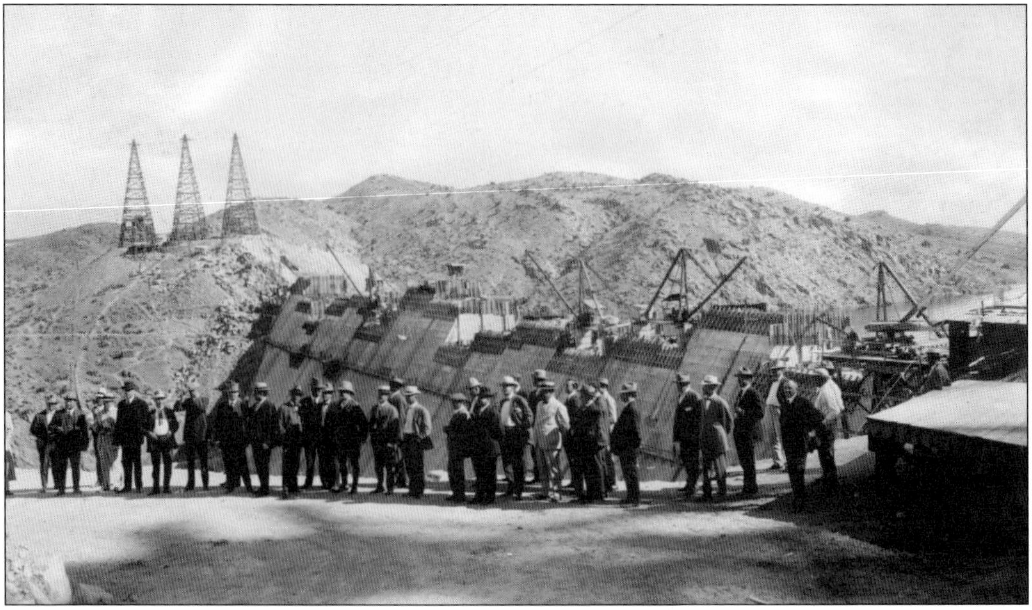

The photograph above shows congressmen not only touring the damsite but pausing for a photo opportunity to support the dam. The photograph below showcases the construction of this massive concrete dam. Balanced valves would allow service gates to turn out the water for use in the irrigation canals and ditches of the lower valley. Penstock gates would release water for the operation of the power plant as needed. In the *Rio Grande Republic* of January 29, 1915, talk was being made about the "wheels of a mammoth hydroelectric plant" to provide "cheap power" for plumbing the water out to service the farmers. This project demonstrates American innovation at its finest.

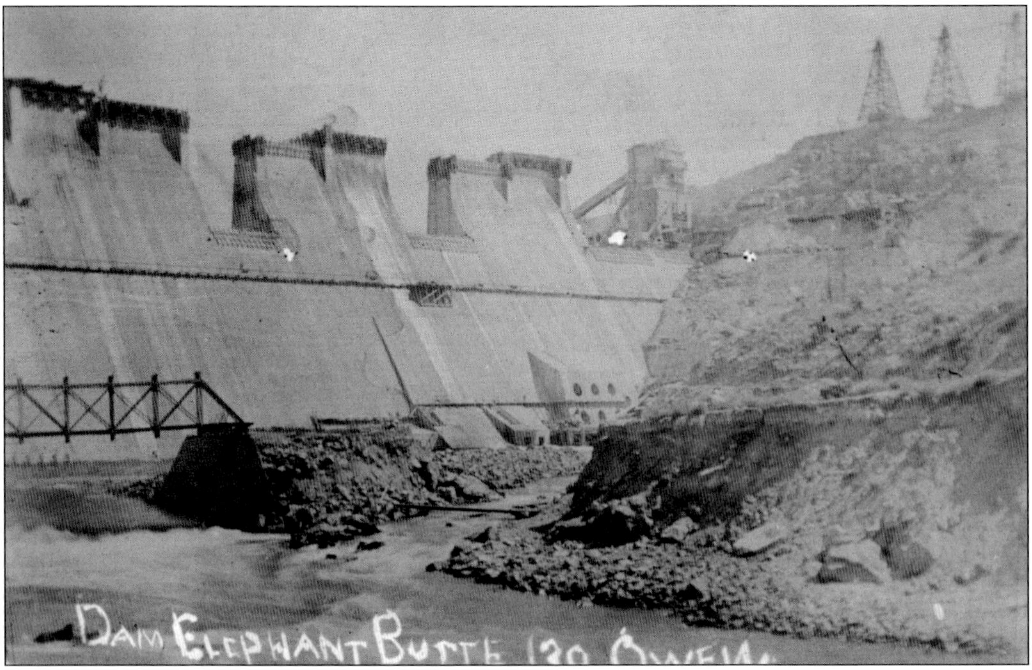

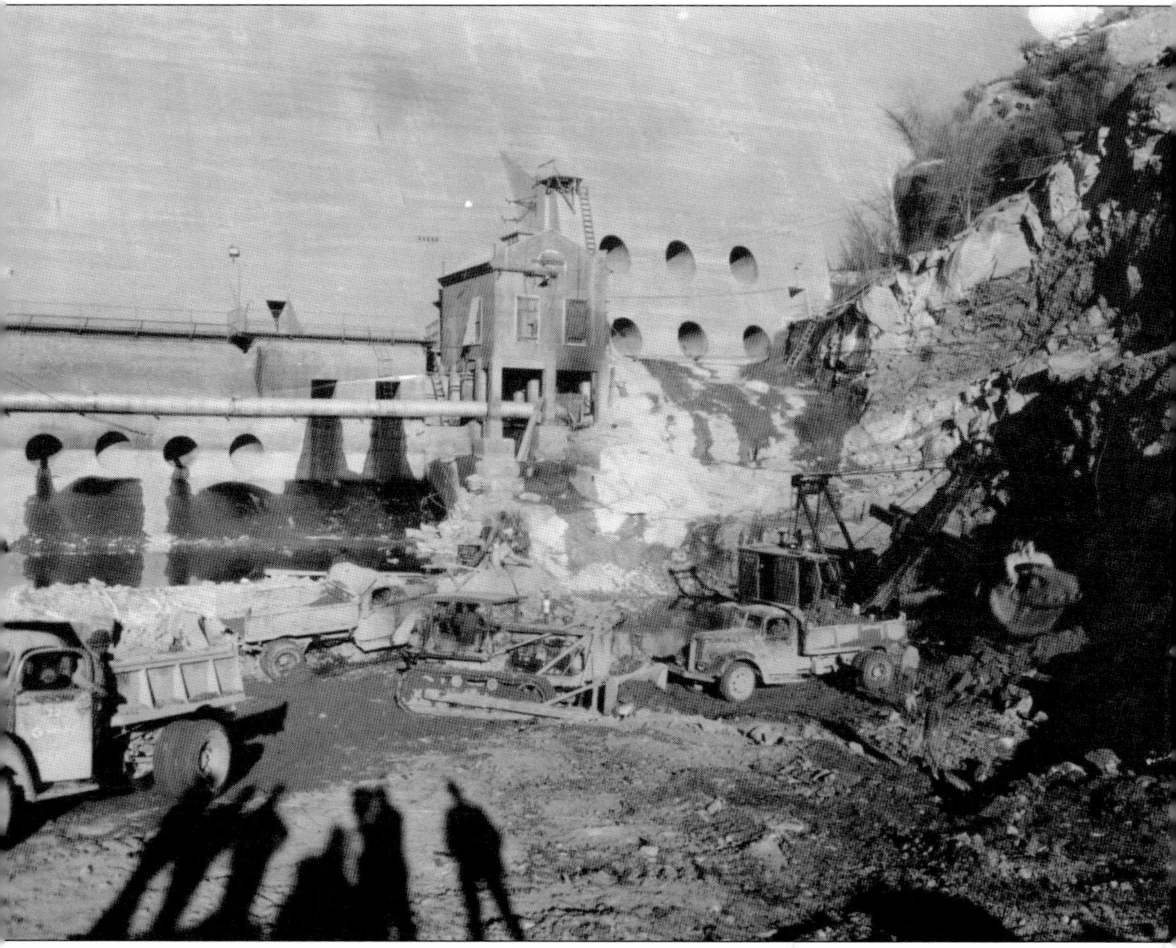

In 1915, newspapers reported that when the artificial lake was filled by the floodwaters of the Rio Grande it would have a 200-mile shoreline. It was estimated that the lake would be 200 feet deep at the dam, and that the average depth would be about 66 feet. Reports estimated the lake would be 1.75-miles wide. According to the January 29, 1915, *Rio Grande Republic*, "Plans are already underway for extensive summer resort features at the lake and at Palomas Hot Springs six miles below. These springs have long been celebrated for their medicinal qualities, and are visited by a large number of people every year." The paper was very positive about the virtues of the lake that was to be built at Elephant Butte. "The big lake will lie in a rugged mountain district, abounding with game and good fishing streams, 120 miles north of the city of El Paso and must inevitably become within a few years a veritable sportsman's paradise."

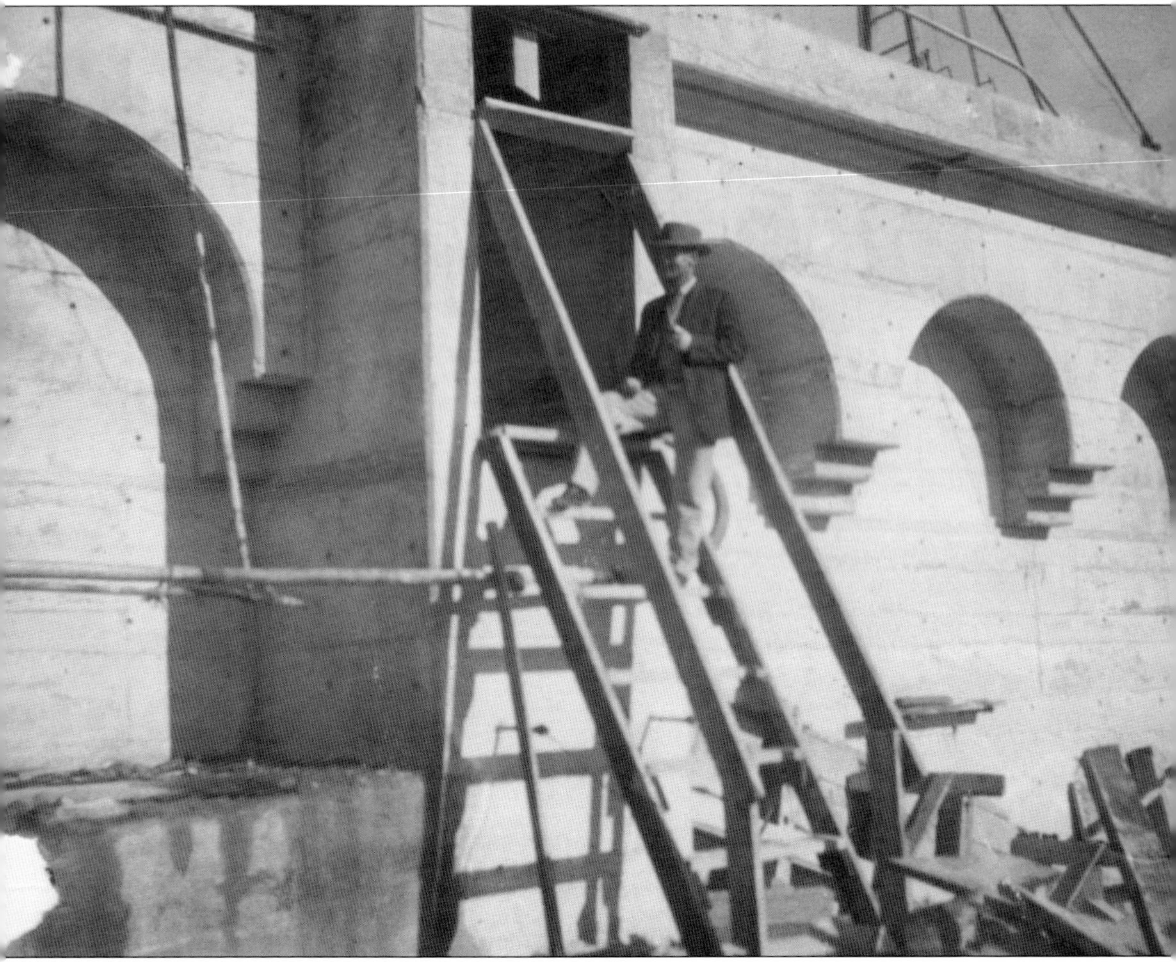

"The up and down stream faces of the dam are cast against forms, giving them a smooth, even surface." The *Rio Grande Republic* of January 29, 1915, reported, "It is difficult, by mere figures, to give an adequate idea of this enormous dam. It is 1,200 feet long on top." The newspaper further listed the stats of the dam as having a maximum height from the deepest point of excavation to the top of the parapet wall at 364.5 feet. The maximum width of the base of the dam was said to be 215 feet, and it was estimated that the road on top of the parapet wall would be 364.5 feet long. Plans were being made in 1915 to stock the lake with game fish within the next few years. "The overwhelming fact connected with it is the transformation which it is making in the Las Palomas, Rincon, Mesilla, and El Paso Valleys," said the paper.

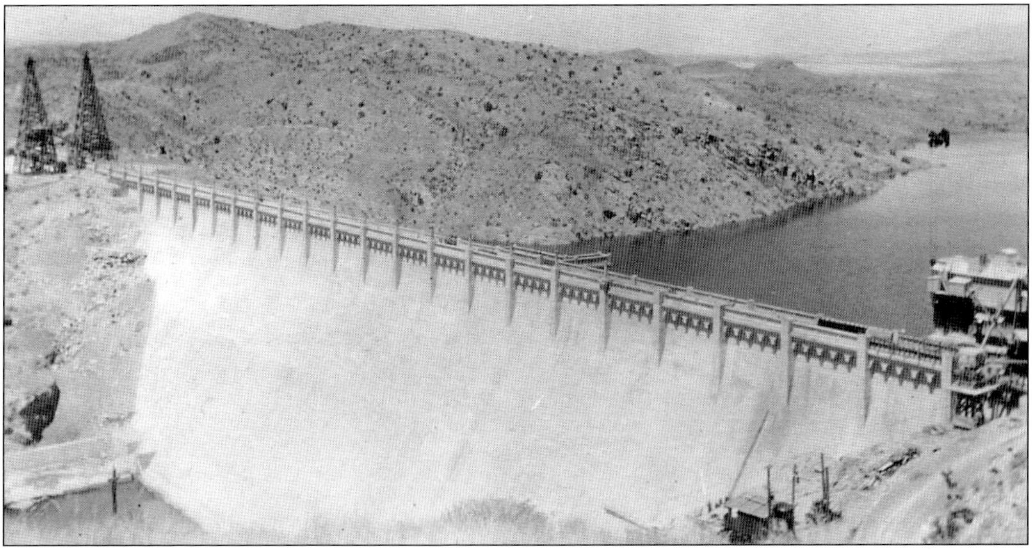

The *Bureau County Tribune* of Princeton, Illinois, from June 2, 1916, announced, "The shrill whistle of the engine at the power plant at Elephant Butte, New Mexico, echoed and re-echoed down the canyon on a recent afternoon at four o'clock. While the same whistle had made similar music innumerable times before during the last five years, its tooting that day had peculiar significance. It gave vociferous and prolonged notice that the last bucket of cement had just been placed in its proper niche in the parapet wall and the greatest storage dam in the world had been finished." It was further noted that "When the floods of the Rio Grande begin to drip over the spillways, the Elephant butte reservoir will contain two-thirds more water than the combined storage of all the reservoirs for Boston and New York."

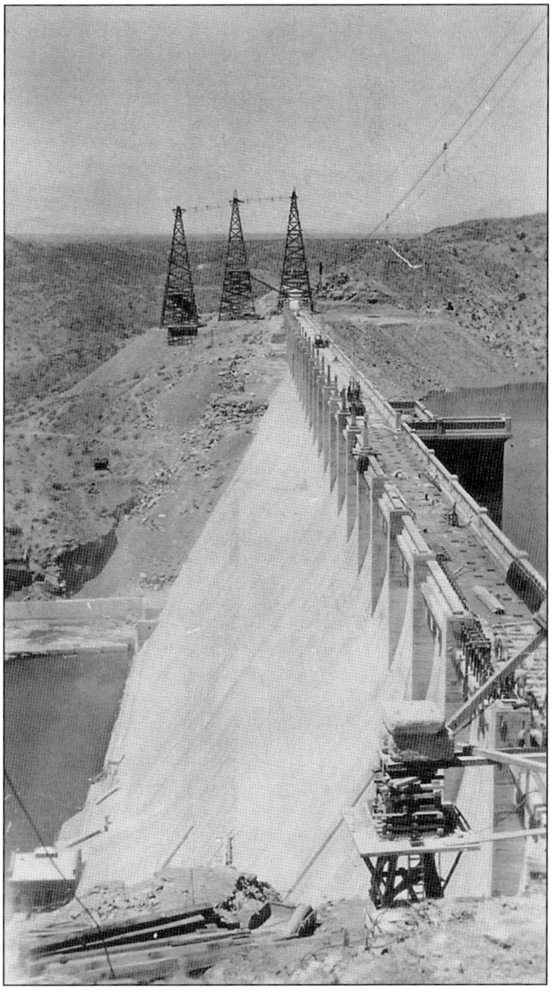

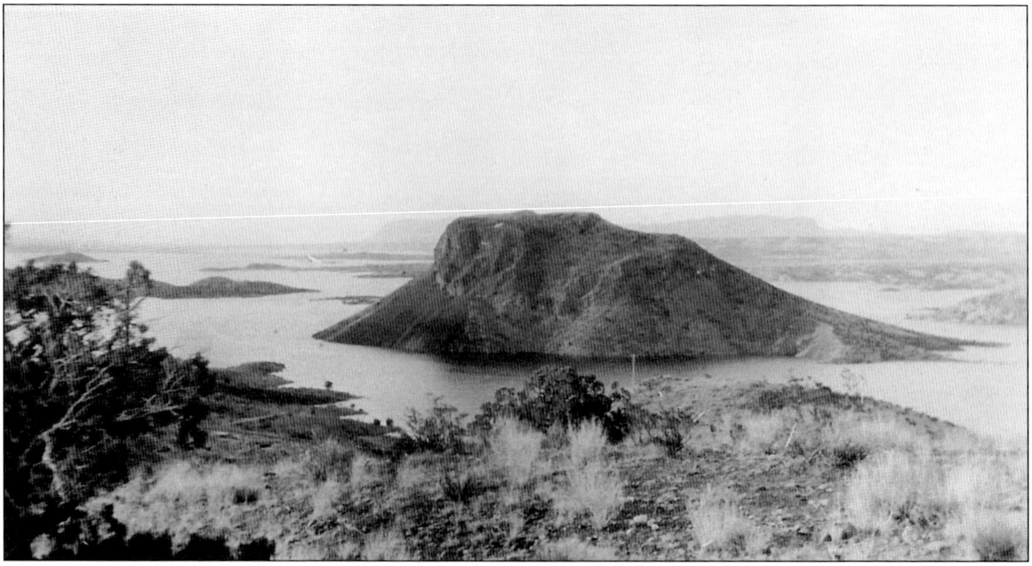

The *Hanover Evening Sun* wrote on October 14, 1916, that Elephant Butte Dam was a "mass of concrete and steel thrown across the Rio Grande river on the desert of New Mexico at the point where a great rock, shaped like a mastodonic elephant, seems to drink from the Rio Grande. This giant impounding dam has formed the largest irrigation reservoir in the world."

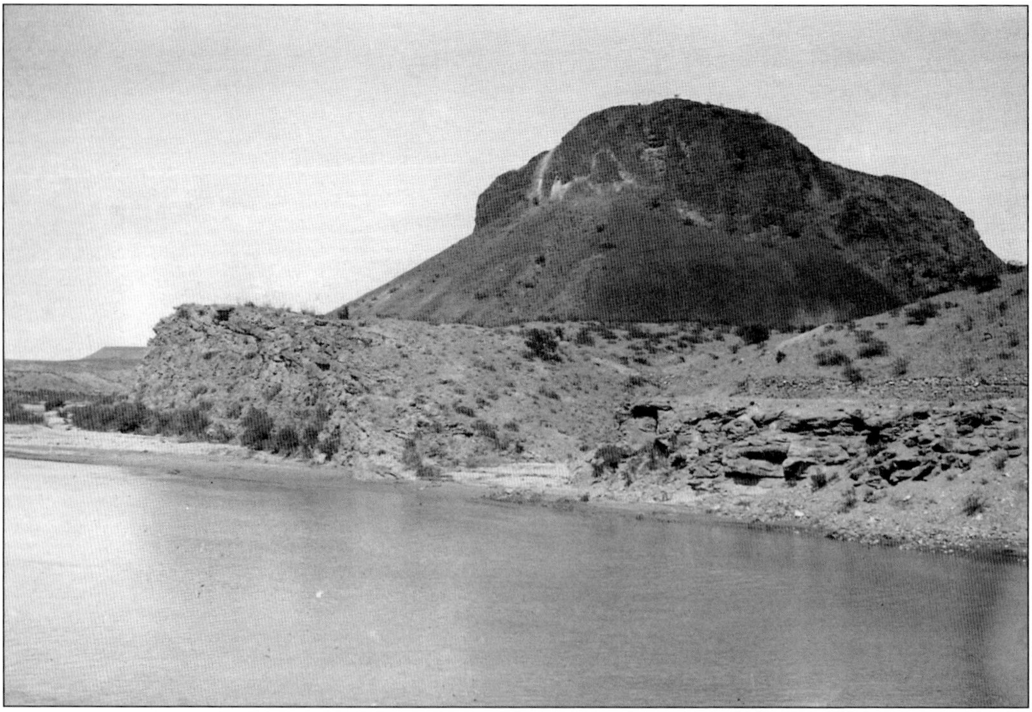

The *Hanover Evening Sun* of October 14, 1916, said that the dedication would "justify the faith of the people of the southwest, who have staked their all on the fertility of the soil and the ability of the project to carry water to the lands when most needed. It will reward the pioneer man and woman of the lower Rio Grande Valleys from Albuquerque to Old Mexico."

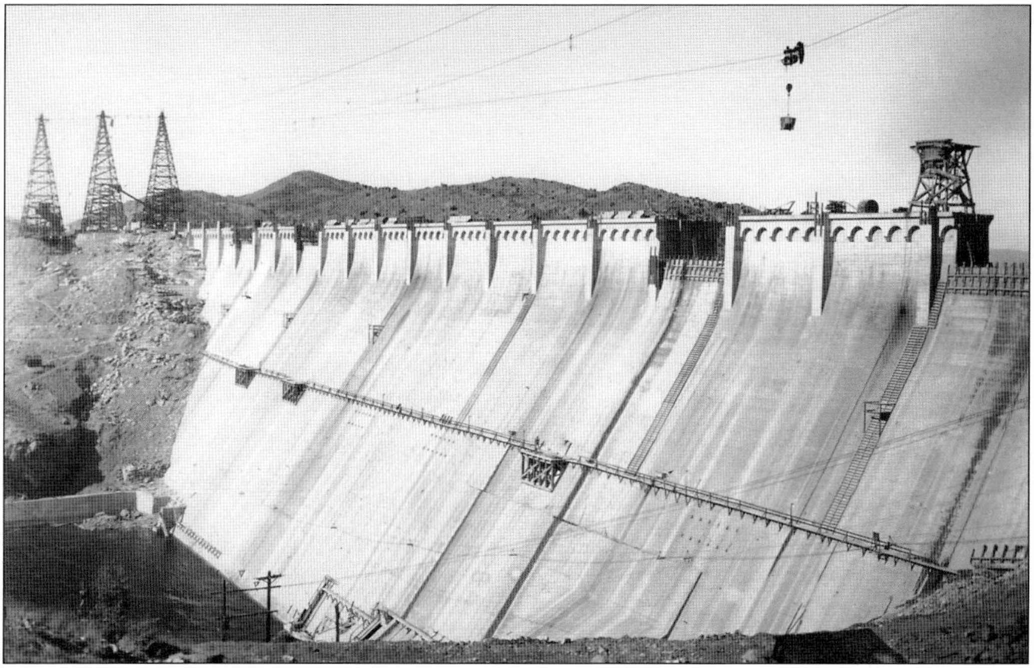

Readers of the *New Castle News* in Pennsylvania learned on June 14, 1916, "About 110,000 acres of land in New Mexico, 45,000 acres in Texas and 25,000 acres in old Mexico will be irrigated" by the waters of Elephant Butte Dam. It was said that the masonry had been placed at a rate of 1,225 cubic yards daily and formed a mass that if placed on a tract of land would be the dimensions of an ordinary city block and rise to a height equal to that of a 13-story building. The gates of the dam had been put into place and the "water in the reservoir stands at thirty-seven feet above the old river bed. When it is filled the average depth will be sixty-six feet."

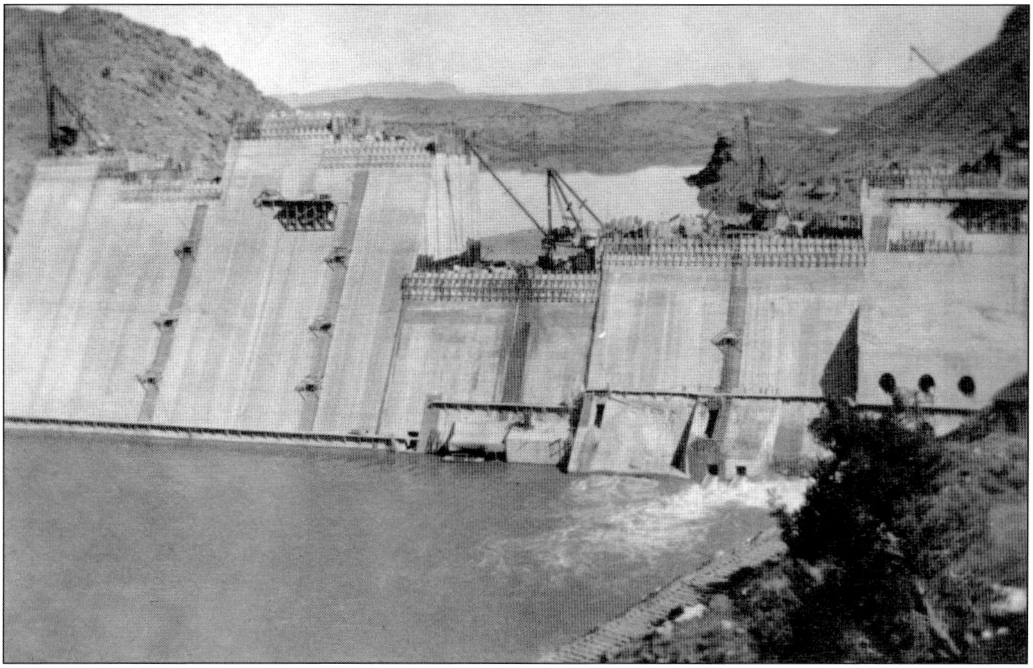

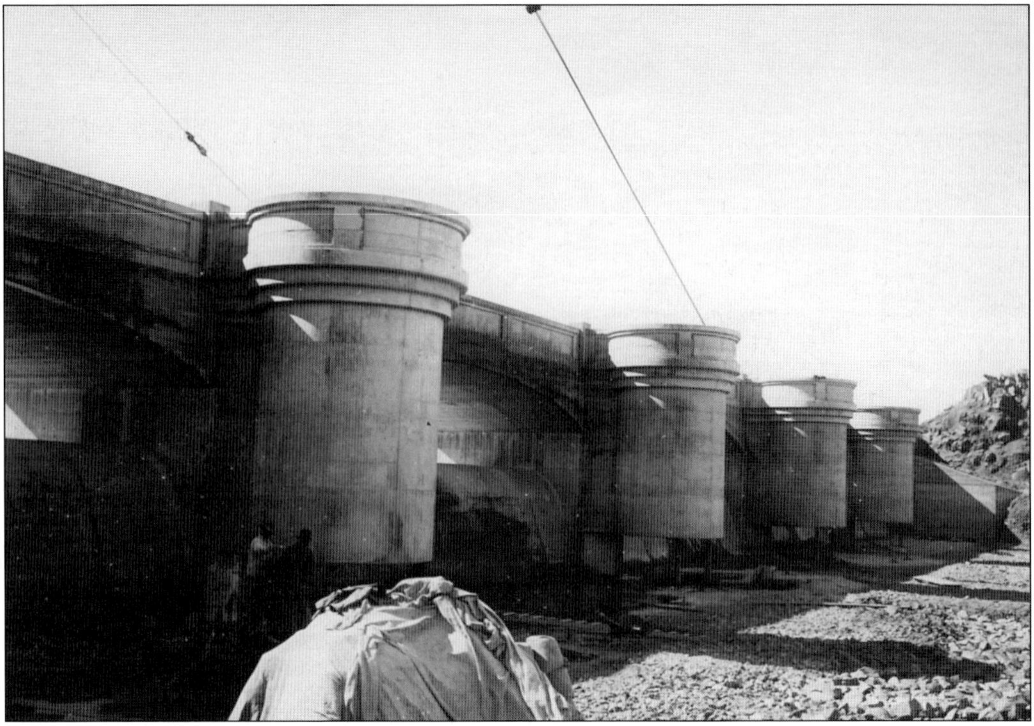

Construction of the spillway was postponed until the height of the stored water was sufficient to be an advantage in excavating the spillway. In order to fit the geological formation of the project, the spillway was designed as a straight channel. It was felt that the water's high velocities resulting from the steep slope would also benefit the design of a straight channel.

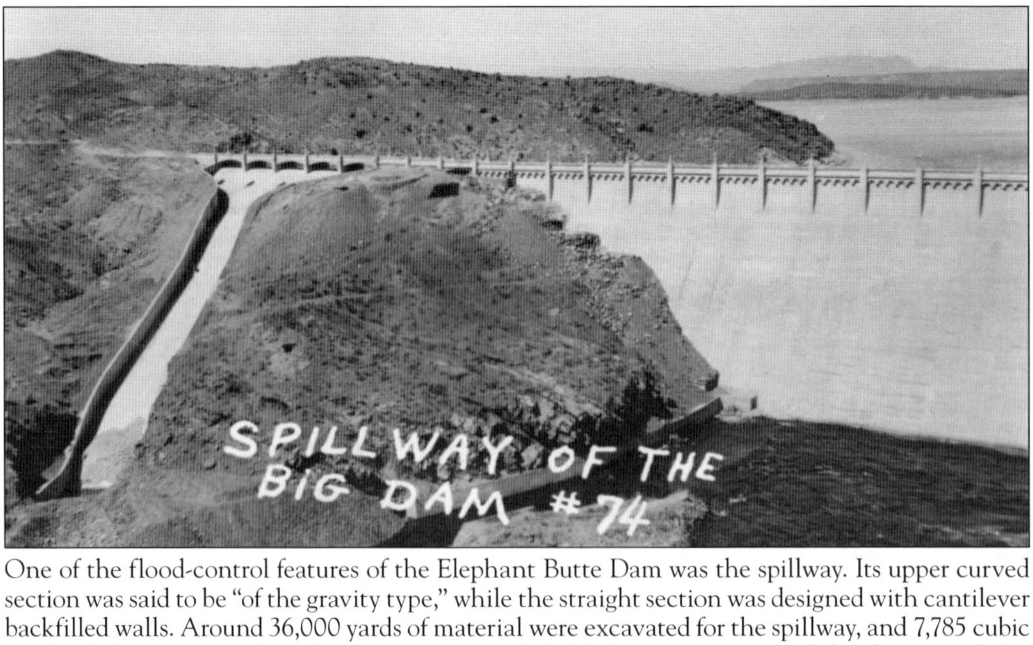

One of the flood-control features of the Elephant Butte Dam was the spillway. Its upper curved section was said to be "of the gravity type," while the straight section was designed with cantilever backfilled walls. Around 36,000 yards of material were excavated for the spillway, and 7,785 cubic yards of plain and reinforced concrete was used. L.M. Lawson supervised the design and the construction of the spillway.

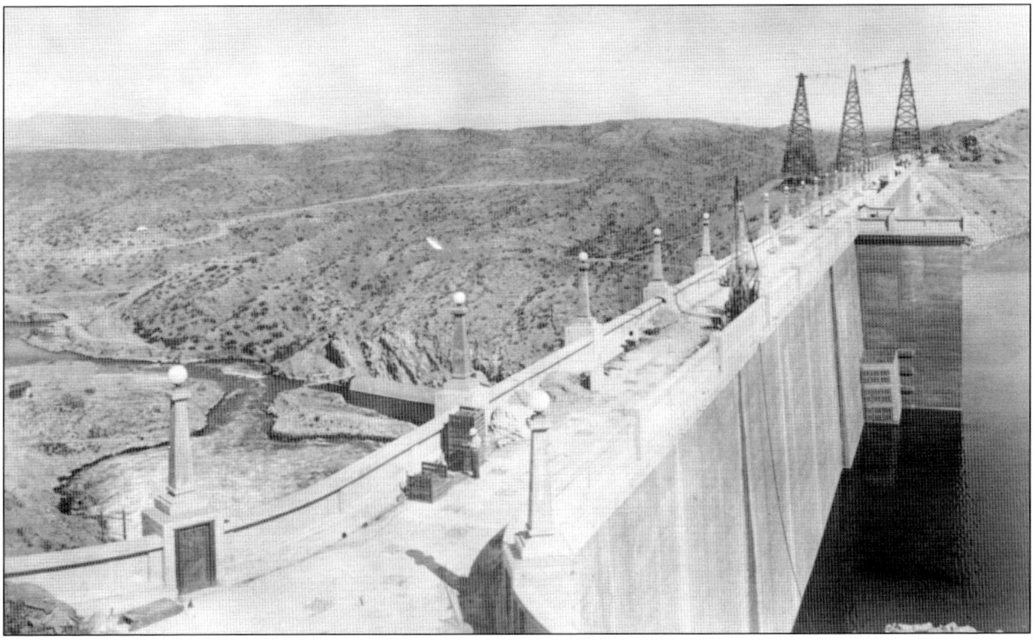

Both photographs show the reservoir of water created by the building of the dam. By 1916, it was still being called Lake B.M. Hall by some newspapers. Of course, the name that would stick would be Elephant Butte Lake. On October 20, 1916, it was reported that a company of Army troops were being sent from Columbus, New Mexico, to guard the dam due to fears that individuals from Mexico were going to try and dynamite the dam. The Reclamation Service had proudly pointed out that it would take a trainload of dynamite to damage the dam. According to the treaty that had been signed with Mexico, the United States would deliver 20,000 feet of water annually to Mexico without cost. In return for the water, Mexico agreed to waive all their rights to the waters of the Rio Grande from New Mexico–Chihuahua line to Fort Quitman, Texas. Of course, the dam was never blown up by any radicals. However, due to the fear of radicals today, the public is not allowed to access the dam.

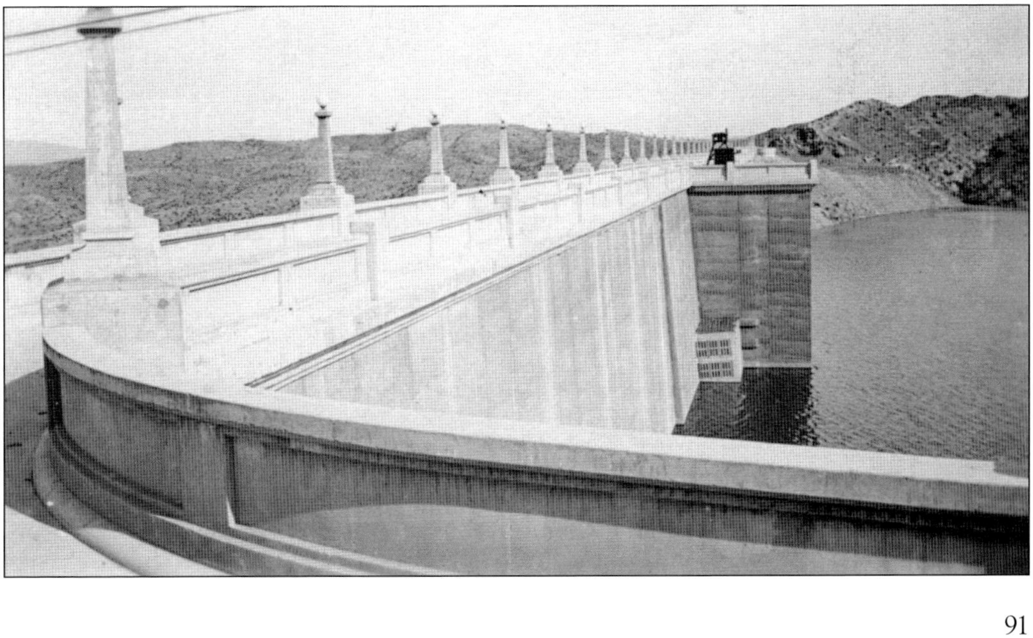

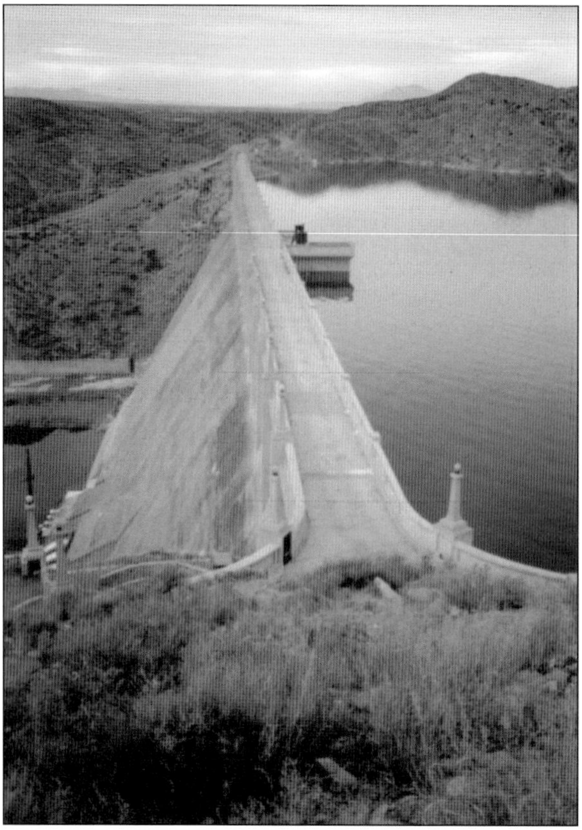

Elephant Butte Dam had been mostly called by that name, but there had been other names attached to the dam, some of which were heatedly disputed. So, what was the actual name for the dam, and who officially, or unofficially, kept changing the name? Interestingly, an article appeared in the *Rio Grande Republic* on October 20, 1916, titled, "Christened the Woodrow Wilson Dam." The story said, "We are met to dedicate to the use of the present and future generations this massive structure and the impounded waters, which, in accordance with the action of the Water User's Associations of Las Cruces and El Paso shall henceforth bear the name of the President of the United States—Woodrow Wilson."

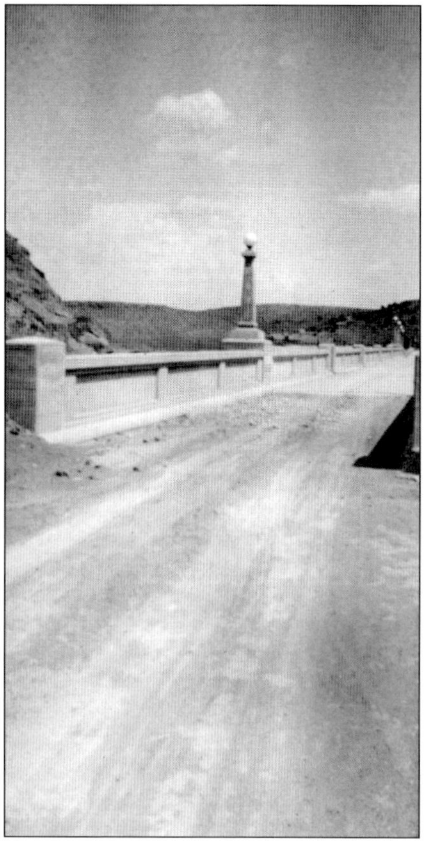

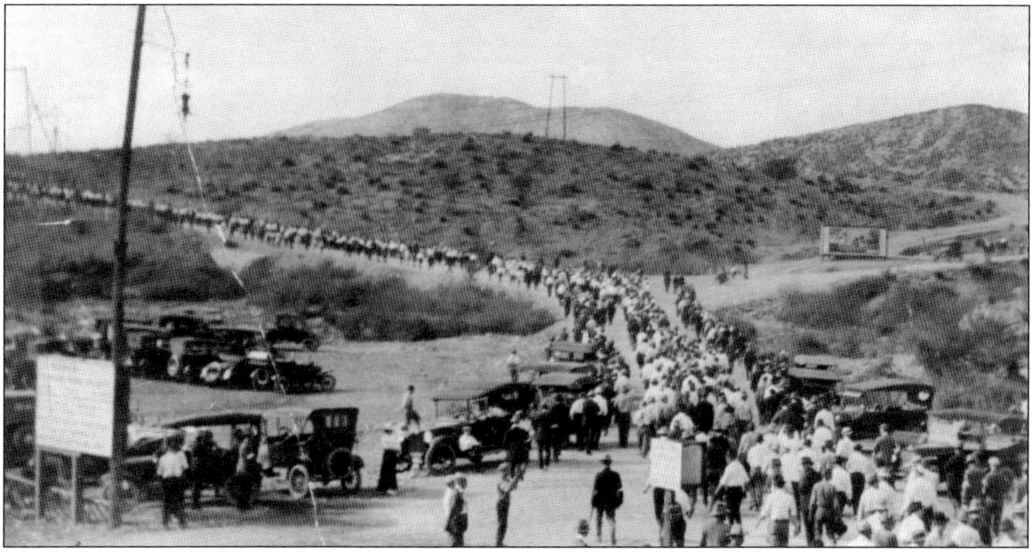

James G. McNary, president of the First National Bank of El Paso, Texas, and chairman of the Board of Control of the Irrigation Congress, was quoted in the *Ardmore Daily Ardmoreite* (Oklahoma) on August 13, 1916, saying, "The dedication of the Elephant Butte dam by President Wilson on Oct. 14, will be an event of world-wide importance in the irrigation field." President Wilson, however, did not attend. A.A. Jones, personal representative of President Wilson, attended along with Brig. Gen. George Bell Jr. and Brig. Gen. G. Morton, commanders of the 10th and 11th Divisions of the US Army. The company and band from the 23rd US Infantry and 350 delegates of the International Irrigation Congress and International Farm Congress, along with other guests, lined the dam and the banks of Elephant Butte Lake.

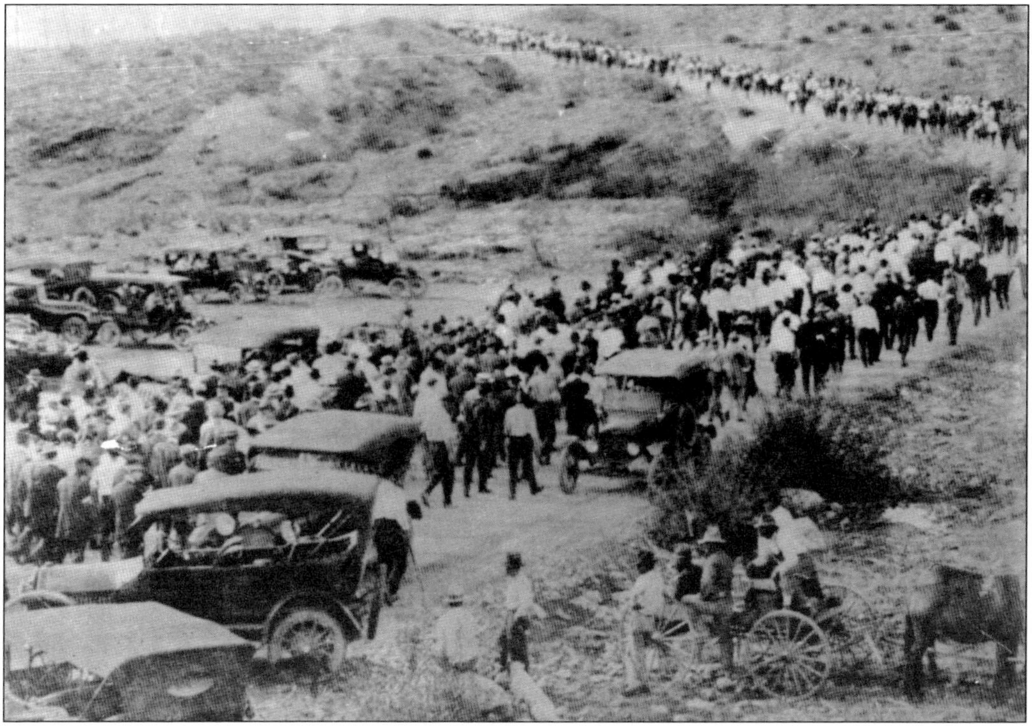

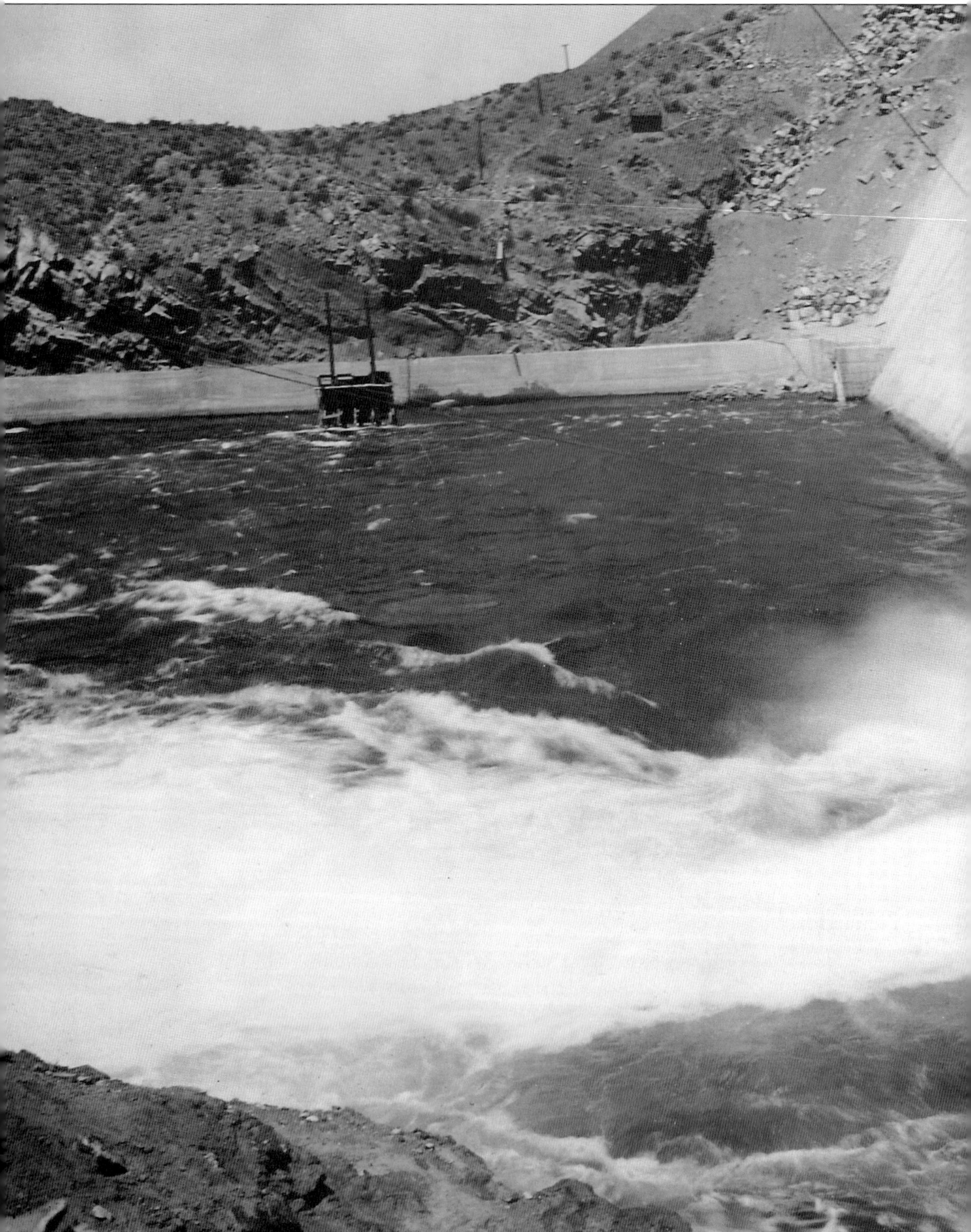

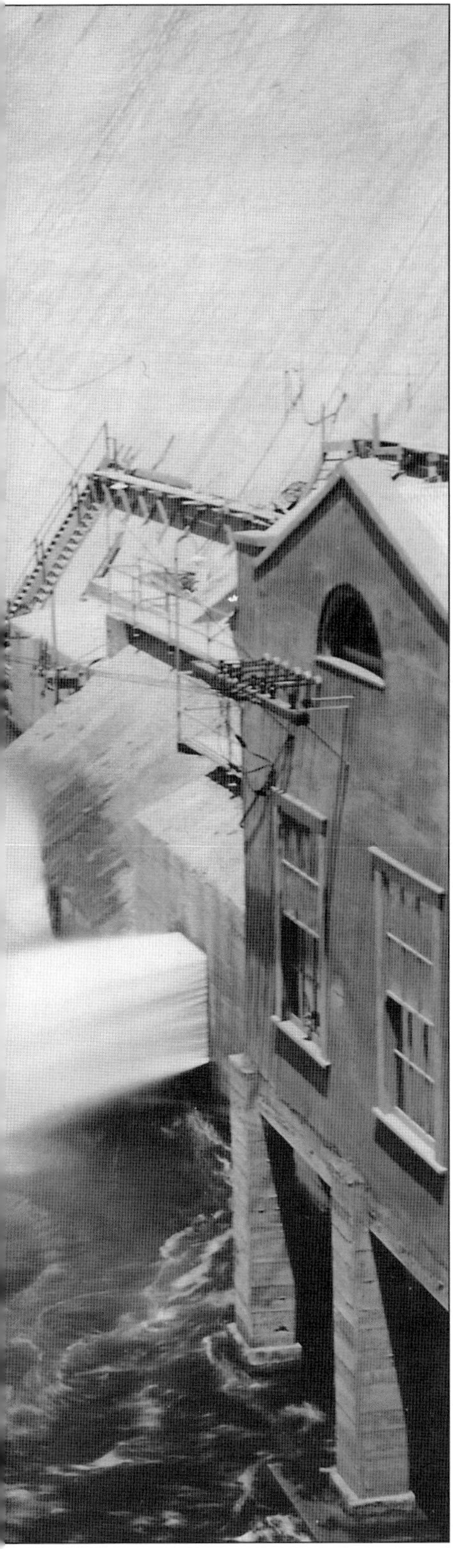

The information that follows is based on the final statistics on the dam, according to the *Syracuse Herald* from October 20, 1916. They said that the dam cost $5 million and had a storage capacity of 862 billion gallons of water, a lake 45 miles long with 200 miles on shoreline, and 610,000 yards of stone and concrete. The length of the dam was 1,674 feet, the width at the base was 215 feet, the width at the crest was 18 feet, and the height of the top above the bedrock was 304.5 feet. The road at the top of the dam was 18 feet long. The actual construction had commenced in July 1910, and therefore it was said that the construction took a span of six years to build. The actual completion date was given as May 12, 1916.

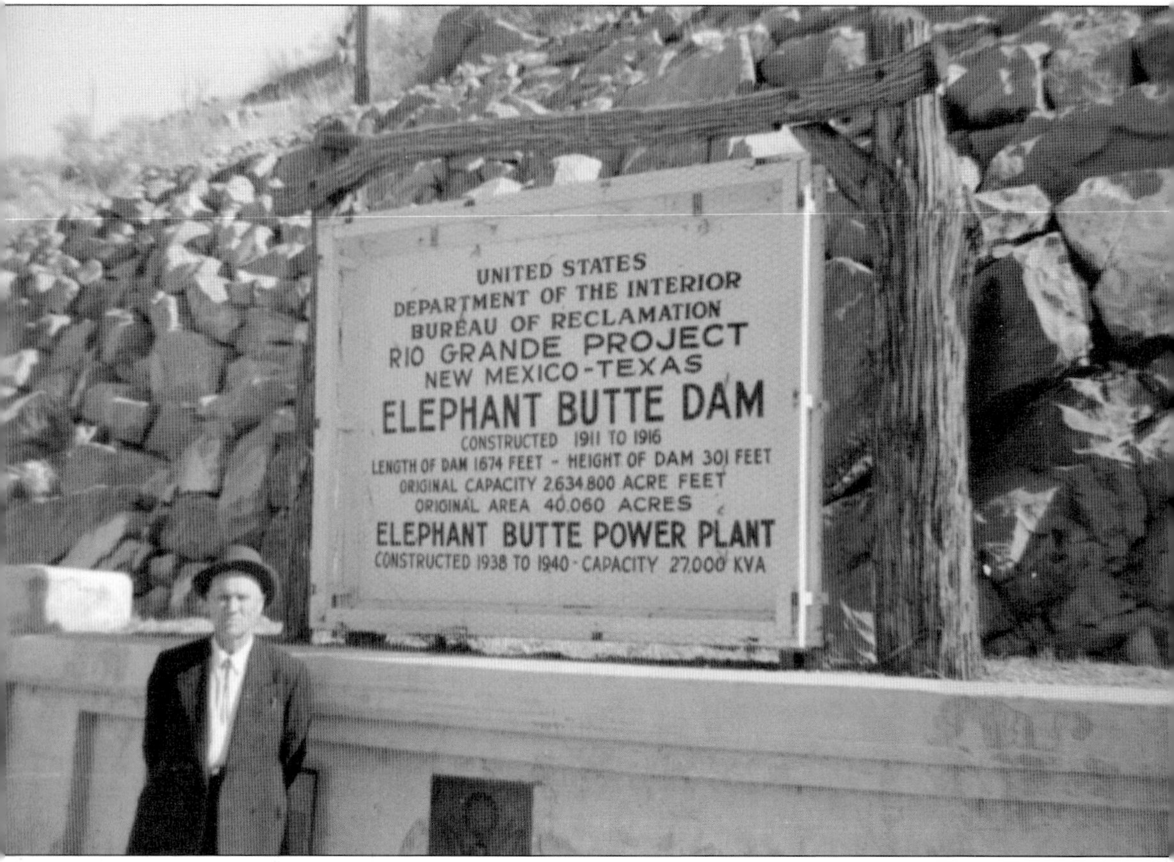

On March 15, 1917, the *Santa Fe New Mexican* ran an article titled, "Heavy Guard for Engle Dam." Although the dam was referred to as Engle Dam in the headlines, it was referred further down in the article as the Elephant Butte Dam. Under the headlines ran the caption "Suspected German at Cuchillo [New Mexico] in Sierra County." The article said US troops from Columbus, New Mexico, were being sent to the Elephant Butte Dam near Engle, New Mexico, to guard the dam for fear if the United States went to war with Germany, the dam would be dynamited. American refugees, who had fled from Chihuahua City, Mexico, reported that a Dr. Knopf had at one time offered to raise a brigade of Mexican troops, drilled and led by German officers, to fight against the United States. On March 4, 1917, President Wilson gave his second-term inaugural speech to the nation, and on April 6, 1917, the United States declared war on Germany.

Five

Recreation and Regulations

The State Park

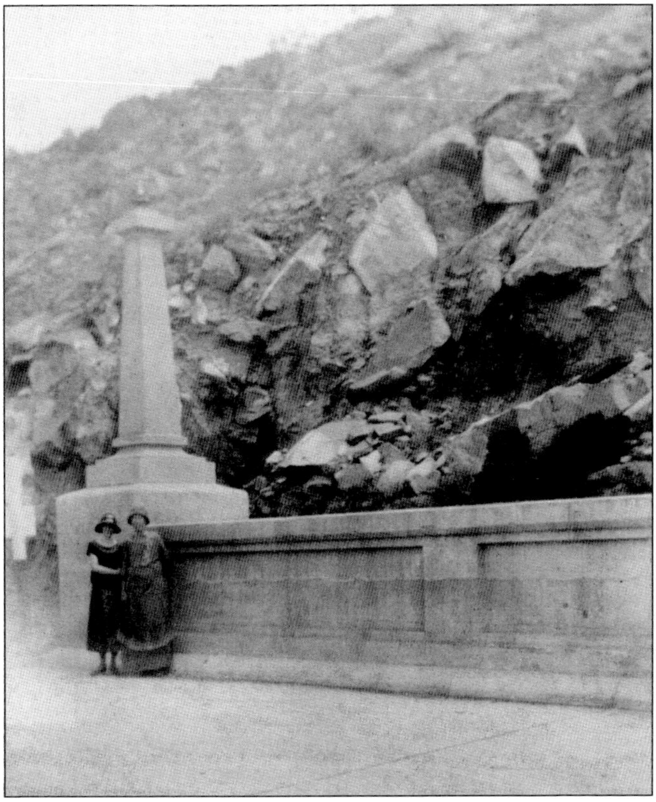

This rare photograph shows a woman with her mother-in-law (notice the difference in the lengths of their dresses) on the road on top of Elephant Butte Dam. Harriet "Hattie" Obedience Pennington, from Coryell County, Texas, was visiting her daughter-in-law Katie Pennington in Las Cruces, New Mexico, and wanted to see the dam, as she had been following its construction in the newspapers.

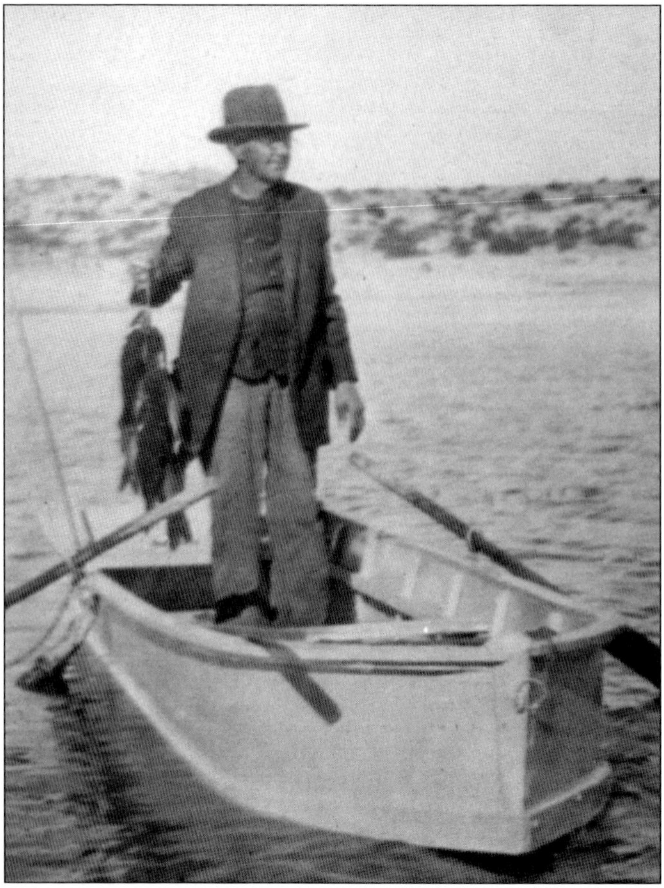

Pictured on the left is local fisherman Thomas Eskew (great-grandfather of one of the authors) fishing at Elephant Butte Lake. The photograph below features one of the earliest boating concessions at the lake. In the 1952 *Albuquerque Journal* from March 18, a man by the name of George Sickles said that he held the first and second concessionaires for fishing and boating at Elephant Butte Lake. He told the reporter, and rightfully so, the lake was first called Hall Lake instead of Elephant Butte Lake. It was on November 18, 1904, that engineer B.M. Hall of the US Bureau of Reclamation submitted his final report to the Twelfth National Irrigation Congress in El Paso, Texas, recommending the Elephant Butte site. Many individuals felt the lake should be Hall Lake in order to honor Hall for his contribution.

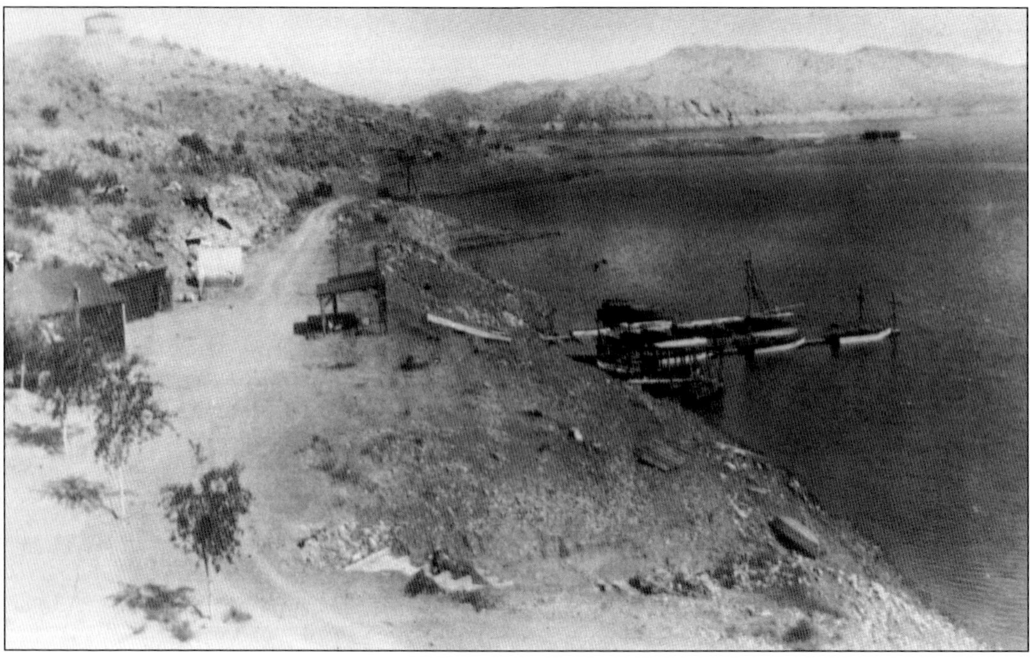

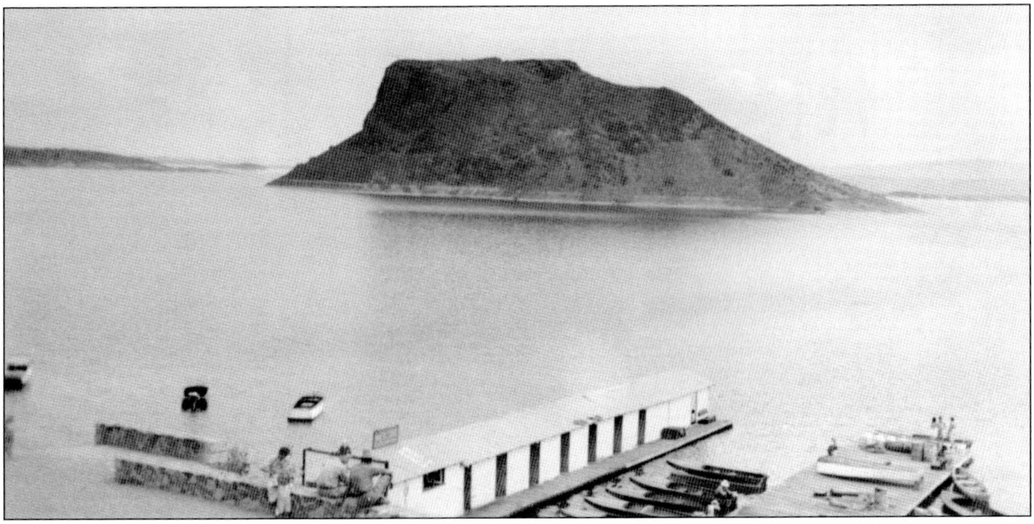

On April 23, 1937, the *Clovis New Mexico Evening News Journal* said the Rio Grande Conservancy wanted to adapt fees similar to those "at government owned Elephant Butte lake, where one can register any size boat for $1.50 per year. Sportsmen who keep boats there, however, usually pay the concessionaire $2 per month for taking care of their craft, so he gets a good deal of revenue from this source."

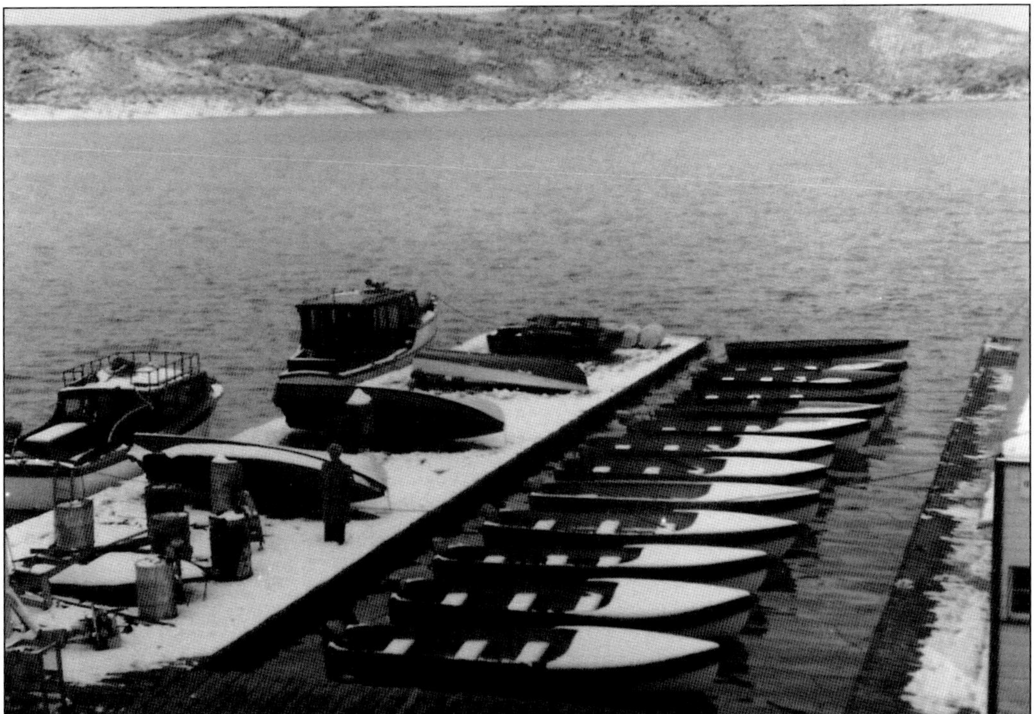

Paul Woolford held the concessions on the marina at Elephant Butte Lake from 1949 through 1954. H.W. and Ida Graves held the concession for the lodge, cabins, and restaurant from 1942 until 1954. Later concessionaires included Kenneth Johnston, Dr. and Mrs. Clark, and Phil Hurdendorf. (Courtesy of Woolford family.)

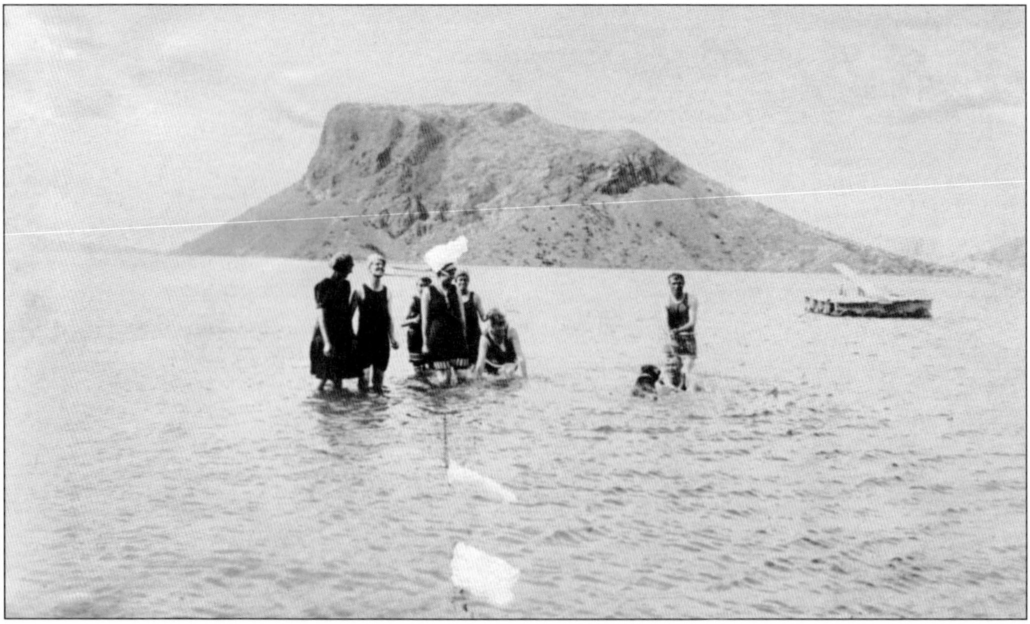

On August 10, 1927, the *Roswell Daily Record* (Roswell, New Mexico) reported a big carnival at Elephant Butte Lake to be held on August 17. It was noted that "attractive prizes have been offered in various swimming and diving events. It is expected that a number of the local swimmers will compete for these cash prizes in the water carnival which is certain to surpass anything heretofore attempted in this time in New Mexico."

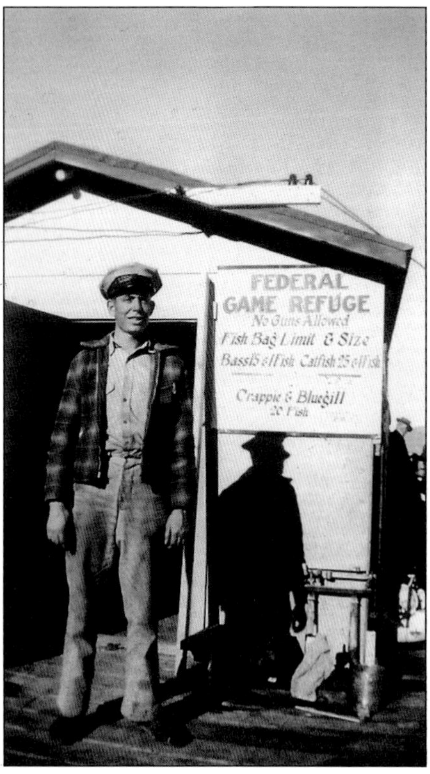

The *Santa Fe New Mexican* on February 10, 1922, reported, "The opposition of [New Mexico] Governor Mechem and State Game Warden Tom Gable to the federal game refuge bill supported by sportsmen all over the country, is baseless and the measure has been entirely misrepresented, according to a letter sent out by Aldo Leopold of the State Game Protective association."

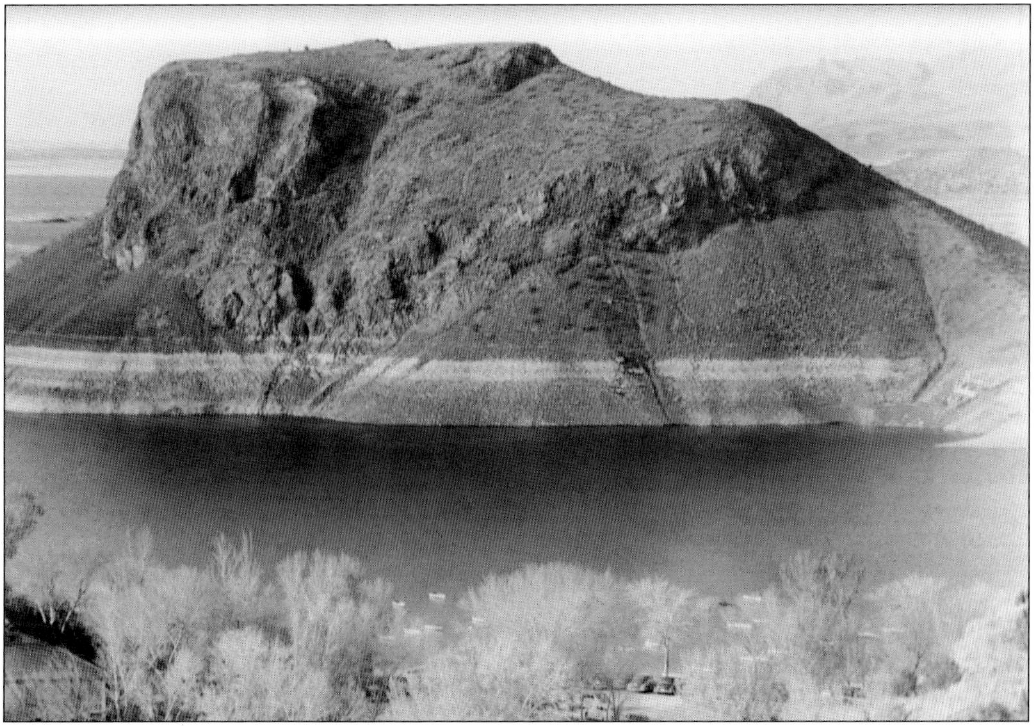

In 1936, boats and motors could be rented or bought at the docks near the dam, Rock Canyon, or the Narrows. Glenn Mims, affectionately known as "Pop" Mims, was expanding the concession to include more boats, bigger boats, and more fishing docks. He had recently purchased a number of new Thompson boats and Johnson motors up for sale or rent. Speedboats were also available. In 1937, Mims, assisted by Jay Sharp, built the docks at the damsite, according to Nita Nelson Mims. She said that Glenn Mims held the concessionaire at Elephant Butte Lake from 1937 to 1943. The photograph above shows the head of the elephant for which the lake was named, and below is one of the early docks at the lake.

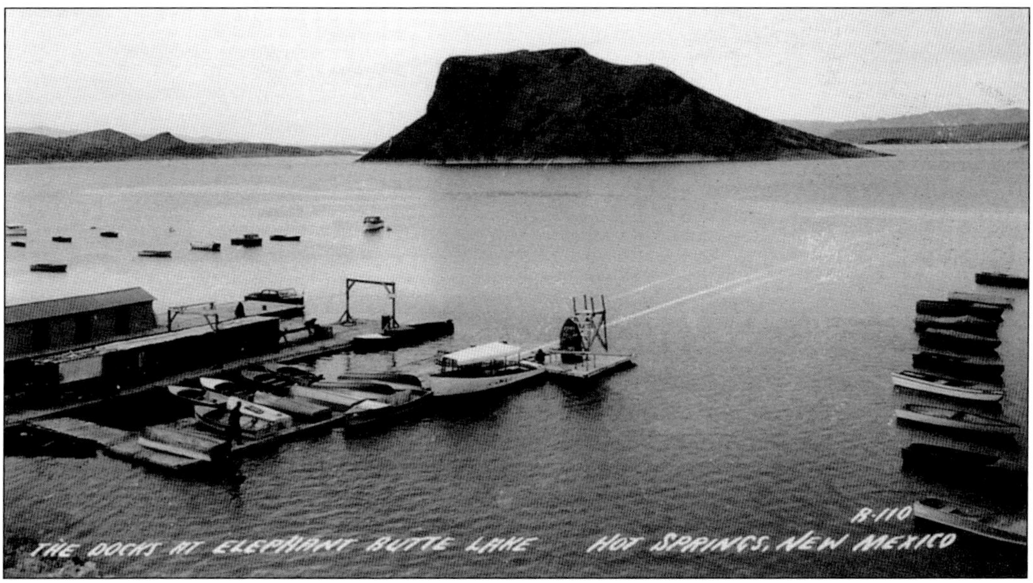

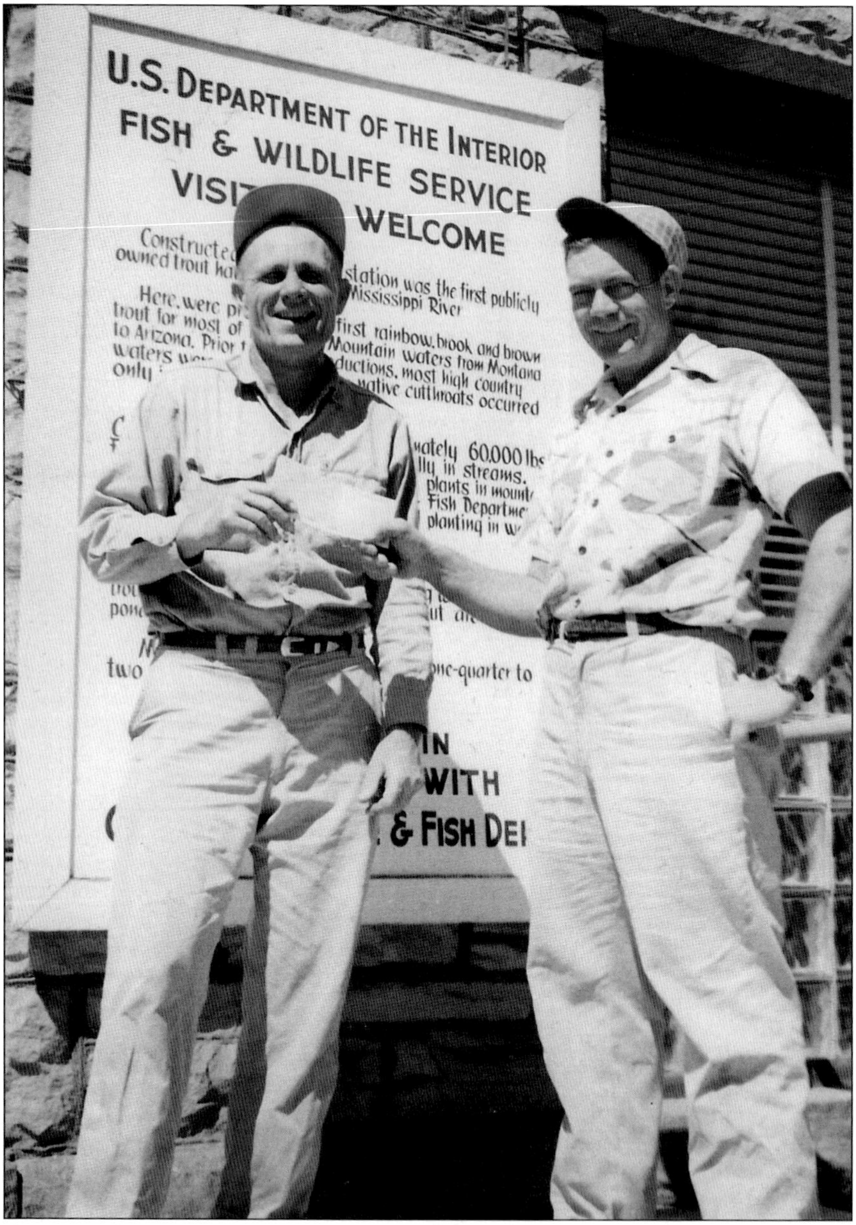

In 1937, the CCC (Civilian Conservation Corps) also constructed a fish hatchery, just below the dam. It was operated by the US Fish and Wildlife Service until 1965, when it became part of New Mexico State Parks. Thomas Nelson was in charge of the construction at the Elephant Butte Federal Fish Hatchery. Residents of Hot Springs, El Paso, Las Cruces, and other surrounding areas had been urging the State Bureau of Fisheries at Elephant Butte to construct a hatchery in order to supply a large number of stocked fish to Elephant Butte and surrounding areas. At one time, it was planned that 25,000 to 30,000 black bass would be moved from the federal hatchery located in Dexter, to Elephant Butte Lake by a railroad tank car. This was the first rail shipment of fish from the new federal hatchery site east of Dexter. The railroad car was known as the "federal fish car." Unfortunately, the shipment was delayed and the fish had to be planted in Tansil Lake near Carlsbad, New Mexico; Black River, New Mexico; and other eastern lakes.

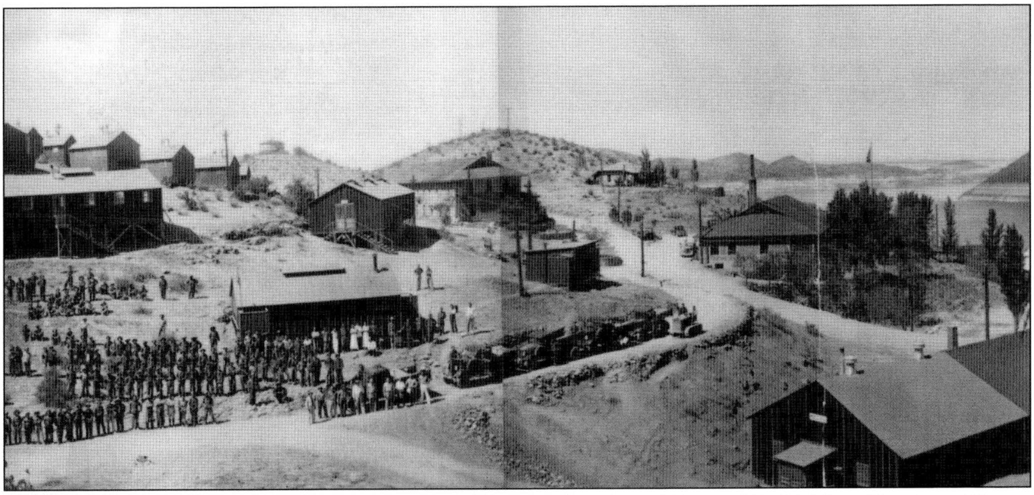

There were two CCC camps, Camp BR-8-N (shown above) and Camp BR-54-N, at Elephant Butte, by 1935. Their goal was to engage in park and recreational work at the south end of the lake. Their work included roads, water supplies, power lines, landscaping (including planting trees and shrubs), campgrounds, fills, boat docks, stone stairway, stone spiral walks, and a boat launch.

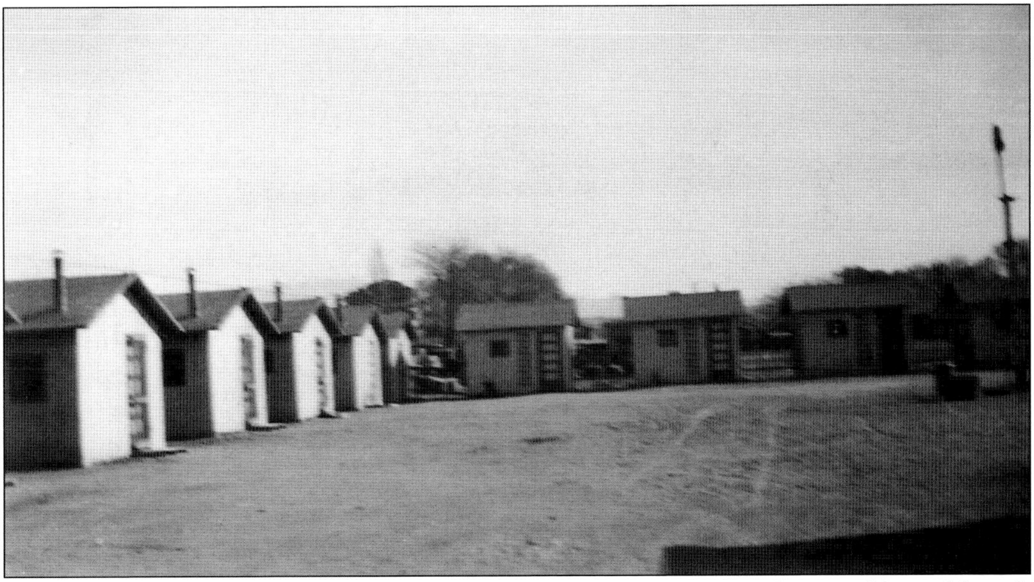

The CCC built 19 tourist cabins, comfort stations, pavilions, fireplaces, table and bench combinations, and tourist camps. In the fall of 1935, the Soil Conservation Service began the deployment of CCC camps to work on conservation measures aimed at reducing siltation in Elephant Butte Lake. By 1937, silt had diminished the reservoir's capacity by 20 percent. The photograph below shows the camp at the south end. (Courtesy of Chuck Martin.)

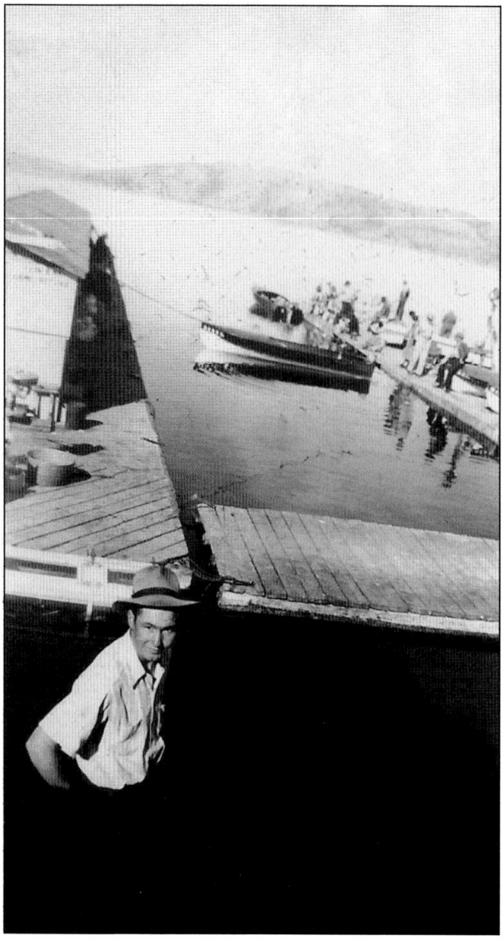

The photograph on the left shows a unique camera perspective of the boat docks at Elephant Butte Dam; boaters came from all over the state to enjoy the waters of Elephant Butte Lake. On April 24, 1942, the *Clovis News Journal* wrote, "Kenneth Mims (shown below), boat dock operator, said three divers from El Paso searched all day Saturday without success for the bodies of A.R. Munoz and Fred Muerdier. They covered a pretty good area, Mims said, and made quite a few dives. But they didn't find anything. They're going to try again Sunday." The El Paso businessmen were fishing with J.E. Bannister when a gale capsized their boat. Bannister's body was found the next day. Standing beside Kenneth Mims is Juanita Nelson, daughter of Thomas Nelson, builder of the Elephant Butte Fish Hatchery.

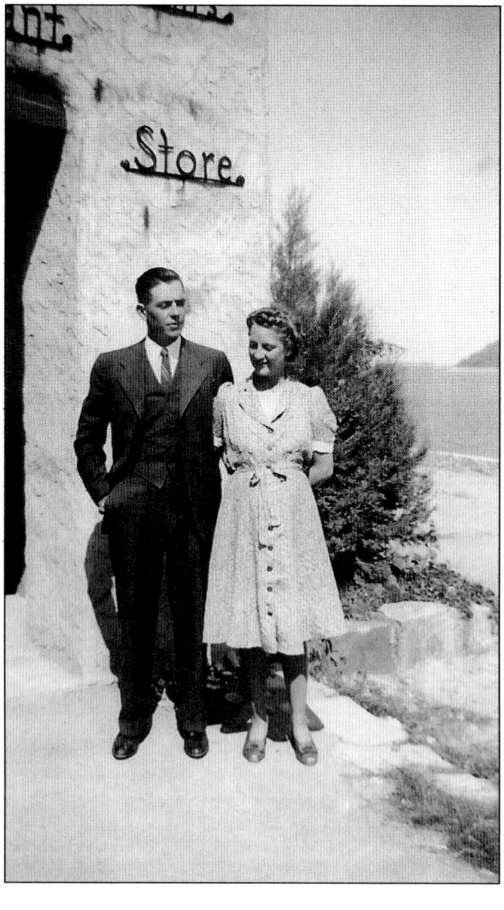

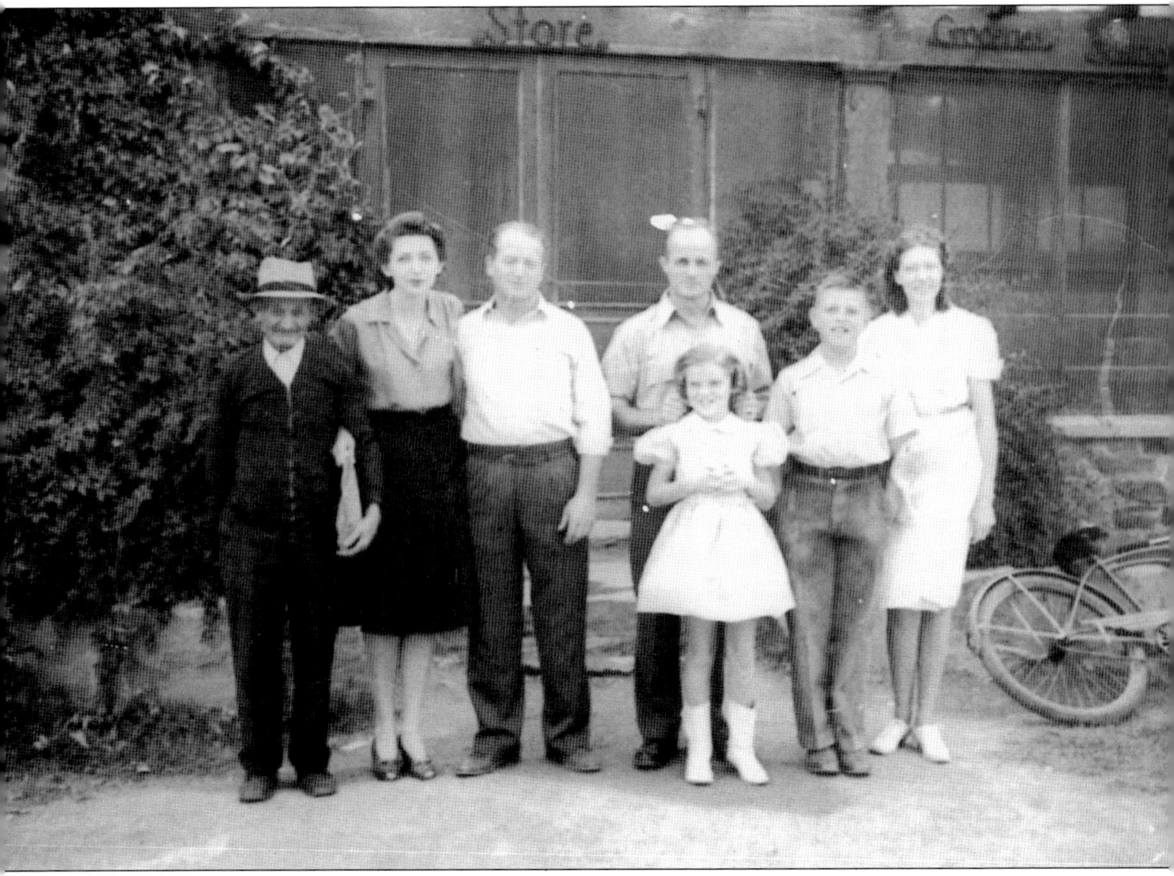

Extended members of the Paul Woolford family pose in front of their concession at the damsite in the late 1940s. The *Lordsburg Liberal* of October 8, 1948, reported that bids were being issued by the Bureau of Reclamation for leasing of concession privileges with buildings and facilities at Elephant Butte on "leases to begin January 9, 1949, and extending for a period of five years subject to renewal annually thereafter for a total period of ten years." Bidders were able to submit bids on any one or more of three schedules. Schedule No. 1 was for the hotel, restaurant, store, tourist cabins, and campgrounds. Schedule No. 2 was for boats and boating facilities for hire to the public, care and servicing of private boats, sale of boats, and sale of accessories and fishing bait. Schedule No. 3 was for "recreational development on the West Shore of Elephant Butte Reservoir by providing boats and boating facilities, and facilities for providing overnight accommodations and serving meals, to the public." (Courtesy of the Woolford family.)

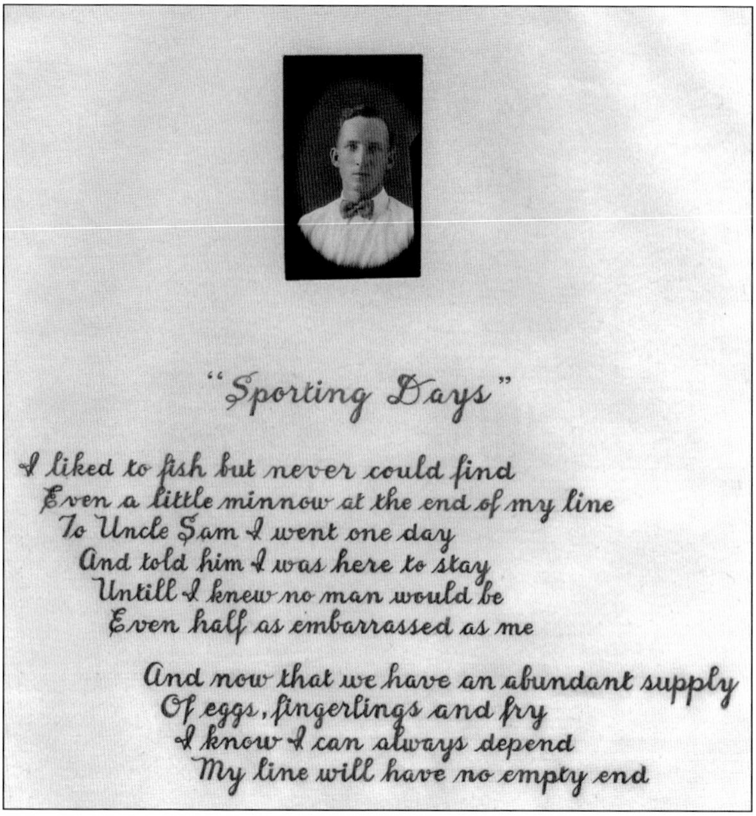

"Sporting Days"

I liked to fish but never could find
Even a little minnow at the end of my line
To Uncle Sam I went one day
And told him I was here to stay
Untill I knew no man would be
Even half as embarrassed as me

And now that we have an abundant supply
Of eggs, fingerlings and fry
I know I can always depend
My line will have no empty end

This poem was written to honor the man pictured, Thomas F. Nelson, superintendent of the federal fish hatchery east of Dexter and later director at the fish hatchery at Elephant Butte. In 1941, Nelson reported that he planted 516,531 bass in Elephant Butte Lake. He also planted crappie and bluegill there.

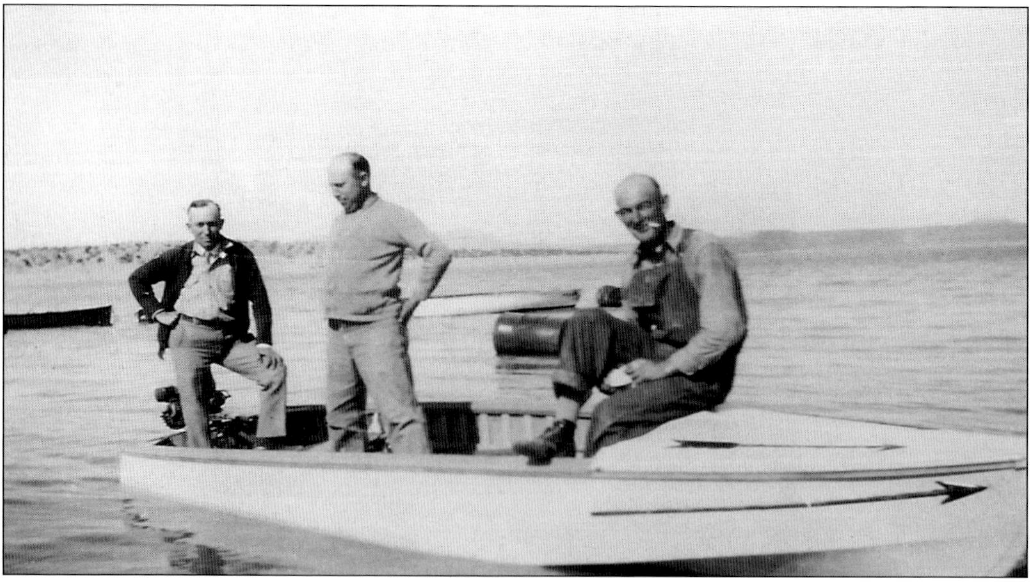

From left to right are Harry Culver, Roy Culver, and Charlie Martin enjoying Elephant Butte Lake in 1932. On March 24, 1932, the *Albuquerque Journal* reported that "preparations are being made here for the opening of the Elephant Butte season April 1. The drone of motors tuning up, boats being painted, stores making displays of fishing equipment make this town a regular sportsman's dream." (Courtesy of Chuck Martin.)

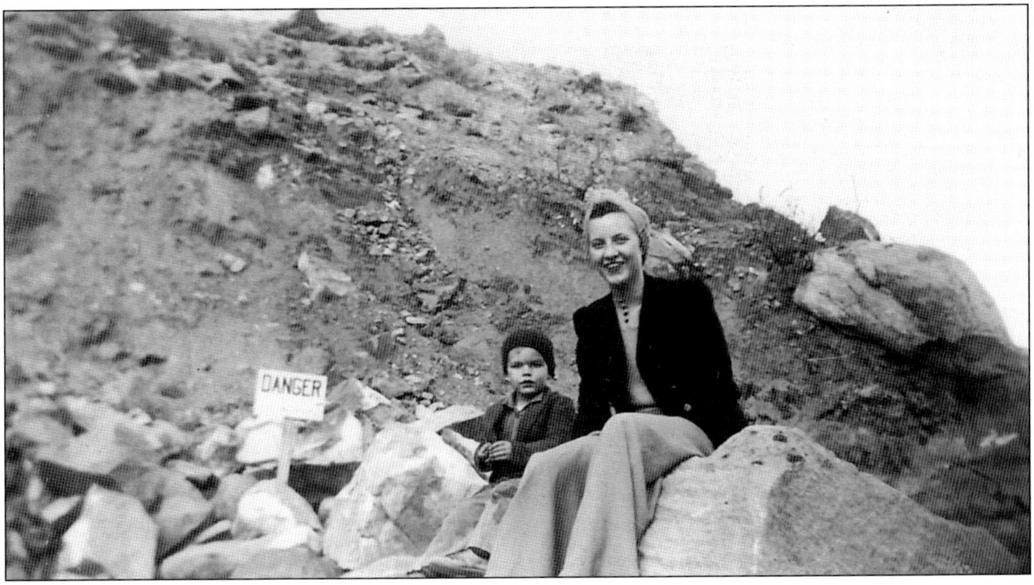

As pictured above, much of the terrain surrounding the lake was solid rock, and every excavation for trees and shrubbery had to be blasted with dirt hauled in and spots leveled. The photograph below shows a water fountain coming from a rock structure. This shows the level of expertise brought to the building site by the CCC workers. Close to this same area, they had planted vines to cover a pergola that provided shade for visitors to the lake. Not seen here is the boathouse that was built at the damsite recreation area. It included space for a locker room, two portal areas (one housed a forge and the other a bathroom), and an oil storage room.

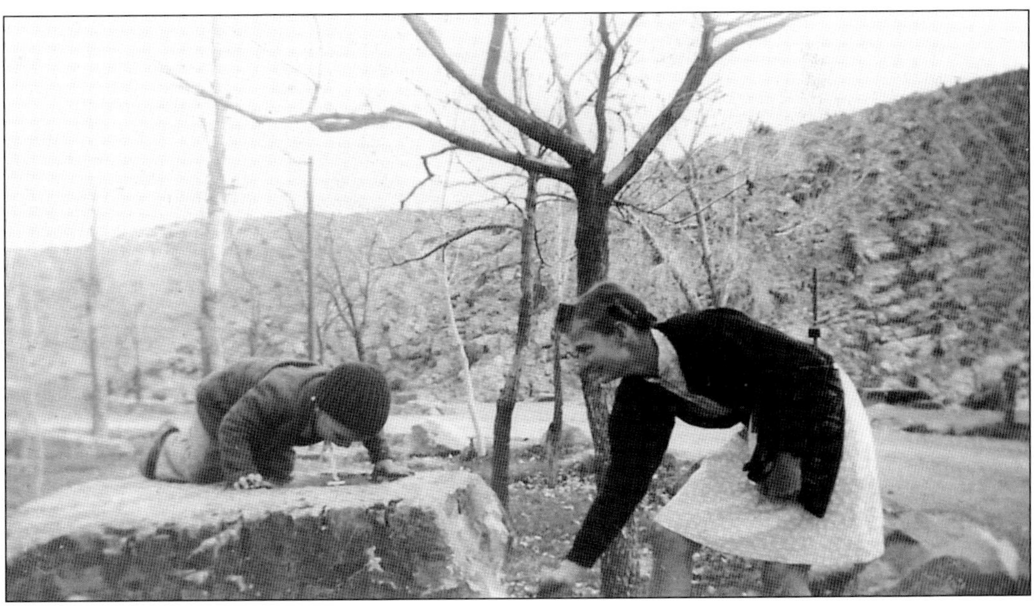

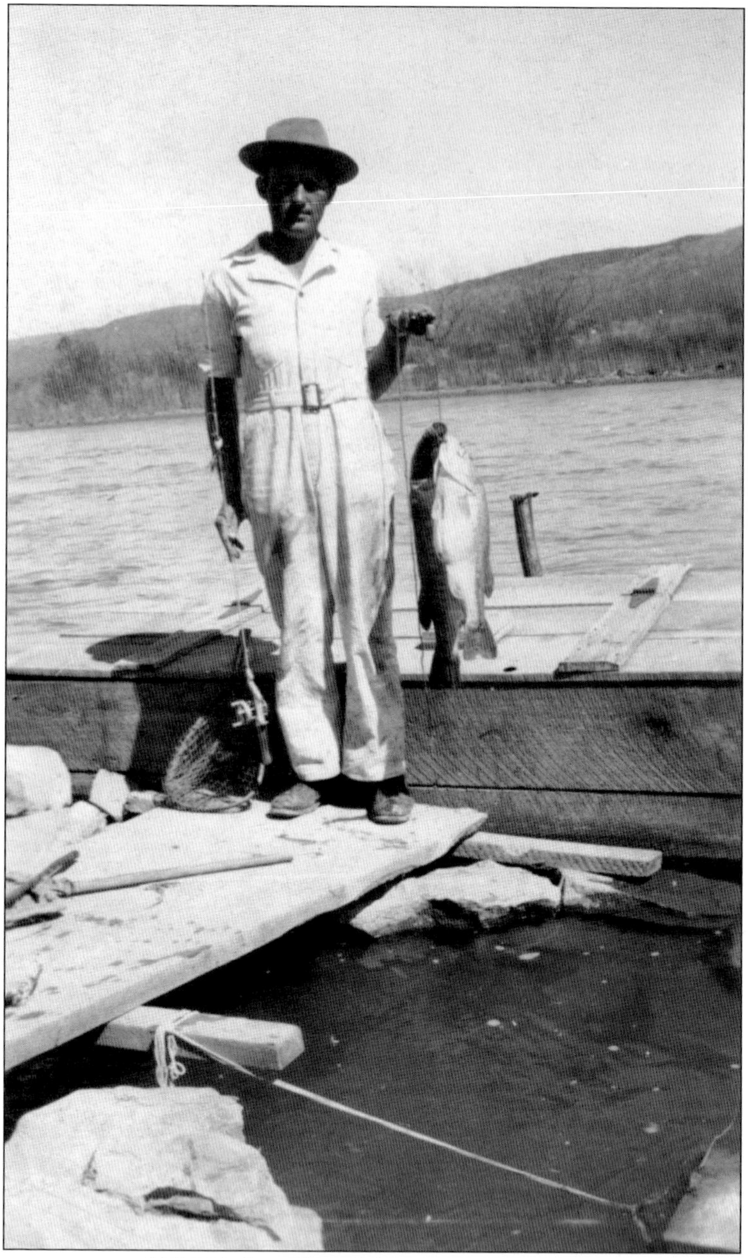

In the 1940s, an average of 70,000 fingerling fish per acre of water were being produced at the Elephant Butte Fish Hatchery. The fish had weighed a total of 589 pounds per acre feet of water and were considered to be "above average" when compared to other fish. The species were black bass, crappie, and bream, and the fish were used for stocking streams and lakes in New Mexico and some of the neighboring states. T.S. Kibbe, head of the fisheries division of the regional office, had said that black bass, crappie, and bream are propagated by selection of best breeding male and female fish and placing them in the hatchery ponds. The male fish build the nests and guard them after the eggs have been deposited until the little fish are spawned. At that time, it was said that there had been such a demand for catfish stock that the US Fish and Wildlife Service would soon begin operating catfish hatcheries at the butte.

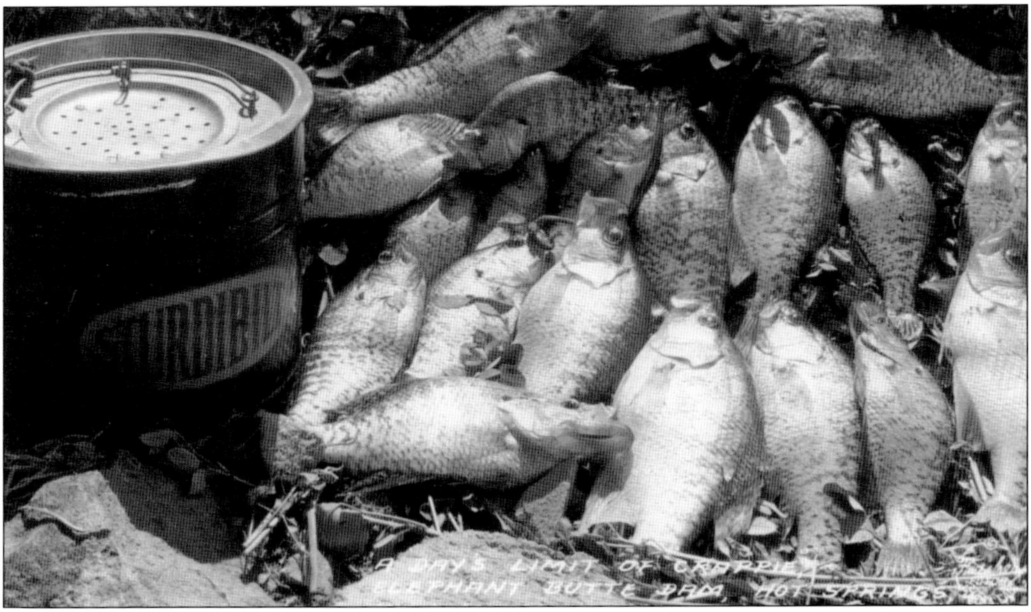

The *Santa Rosa News* reported bag limits on game fish in New Mexico waters on March 12, 1937. For crappie (above), the limit was 20 fish, which had to be a minimum of six inches (Chaves and Eddy Counties had a 10-fish limit). The limit of bass was "15 pounds and one fish. Minimum, 9 inches." Catfish (below) limits were 25 pounds and one fish, minimum eight inches. Ring perch and yellow perch had a bag limit of 40 fish. Bream and other sunfish had a bag limit of 20 fish, and one could catch 10 pike perch per day, but they had to be a minimum of 13 inches. Bass, crappie, bream, and perch season was from April 1 to April 15 and then again from June 1 to November 30.

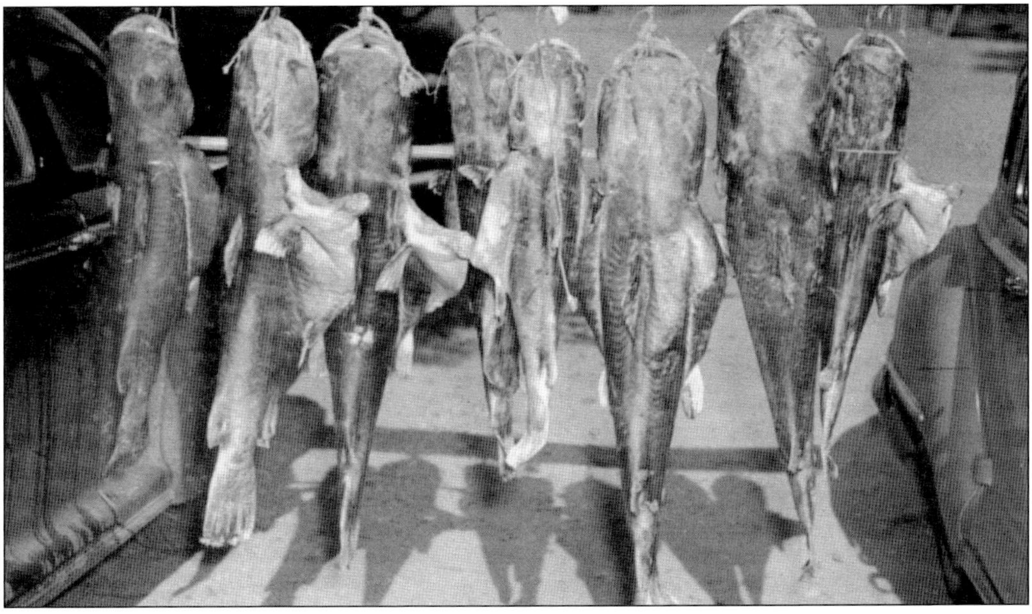

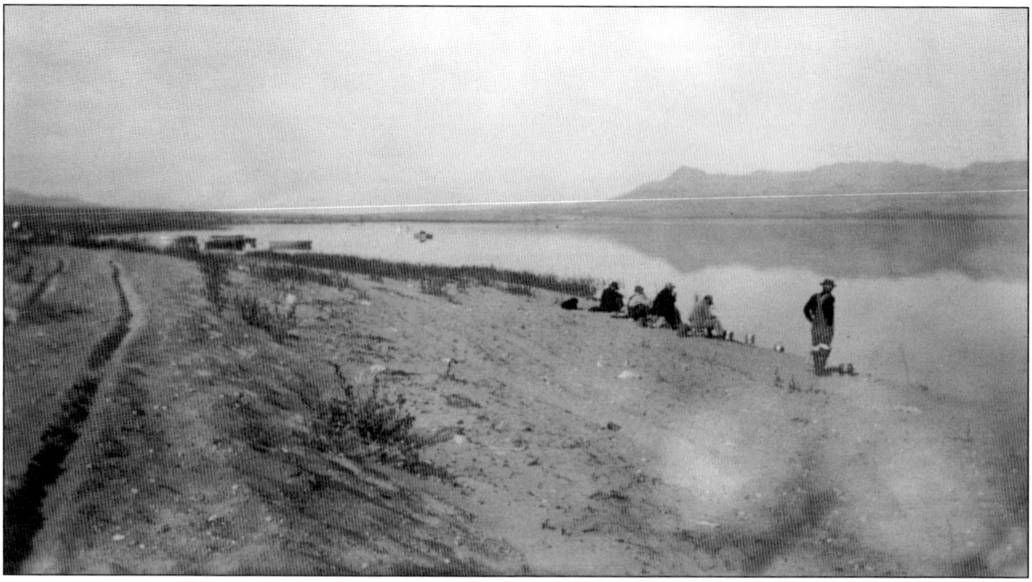

In the early 1920s, the Lions Club cleaned up the beach at Elephant Butte Lake. In honor of their efforts, a portion of the beach was named Lions Beach. The Lions Club was founded by Melvin Jones in 1917 and is a secular nonpolitical service organization. The goal of the Lions Club is to meet the unique needs of their community.

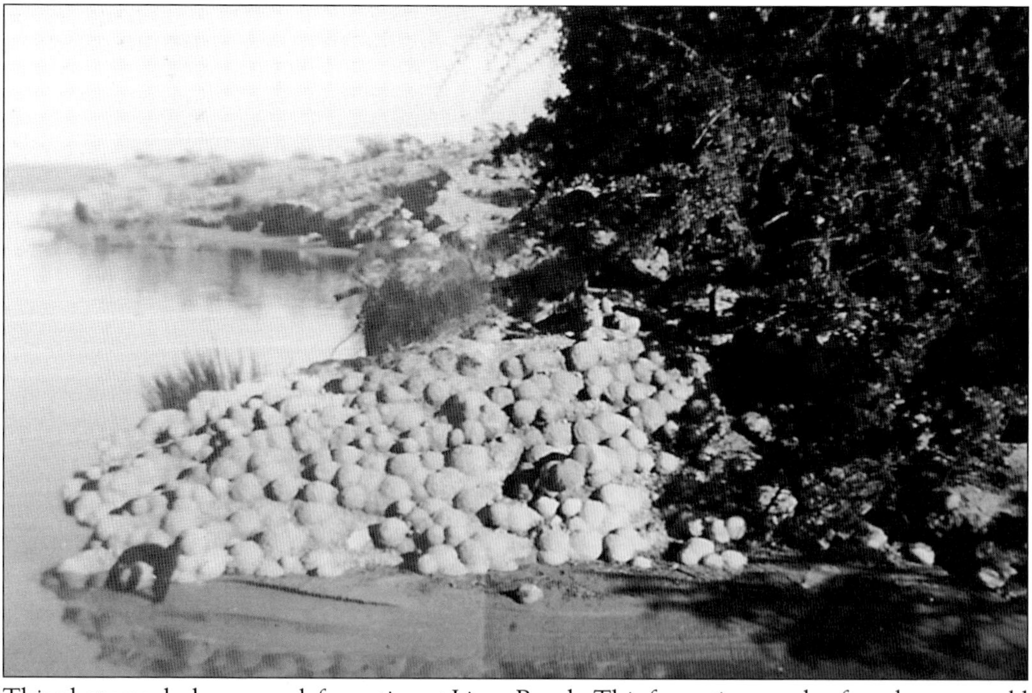

This photograph shows a rock formation at Lions Beach. This formation, made of sandstone, could often be found along the sandy beaches of Elephant Butte Lake. When the lake was down, these types of formations would be exposed. Rock hunters from around Hot Springs would collect the stones and use them in fencing surrounding their homes and businesses.

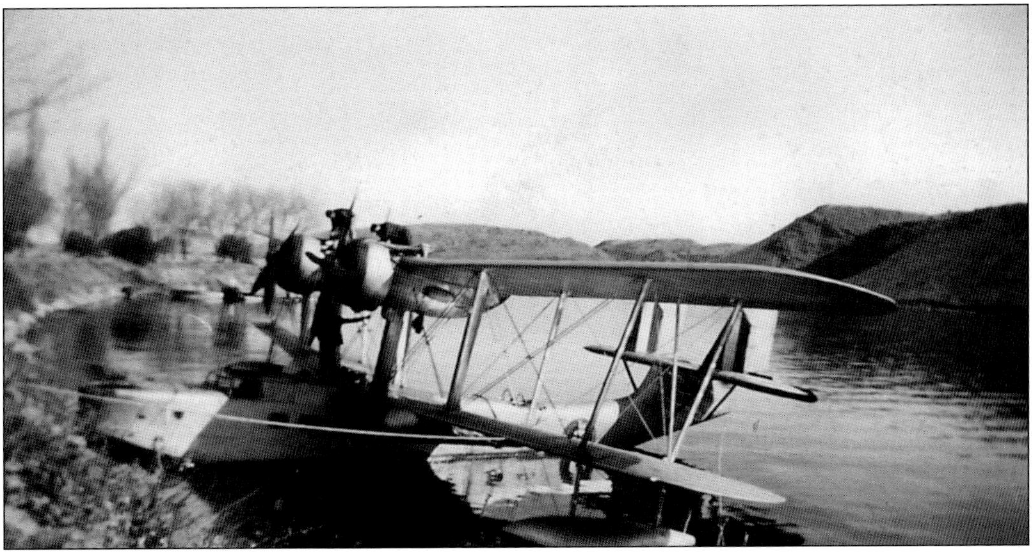

In April 1927, DePenido, an Italian aviator, landed his seaplane (like the one shown above) on Elephant Butte at the invitation of Mayor Leo Smith. He was on a cross-country flight, and after he landed his plane at Elephant Butte Lake he was met at the landing by Hot Springs mayor Leo Smith, the town council, and the chamber of commerce. Schoolchildren were taken to Elephant Butte Lake to watch the landing. DePenido was a guest at a banquet held in his honor at the Vera Hotel in downtown Hot Springs, and there were 42 guests at the banquet, including Mr. and Mrs. Emil James, Mr. and Mrs. J.A. Bullen, Dr. A.C. White, A.J. Howe, and Ross M. Atkins. The aviator soaked in the hot mineral baths in Hot Springs, New Mexico, while there.

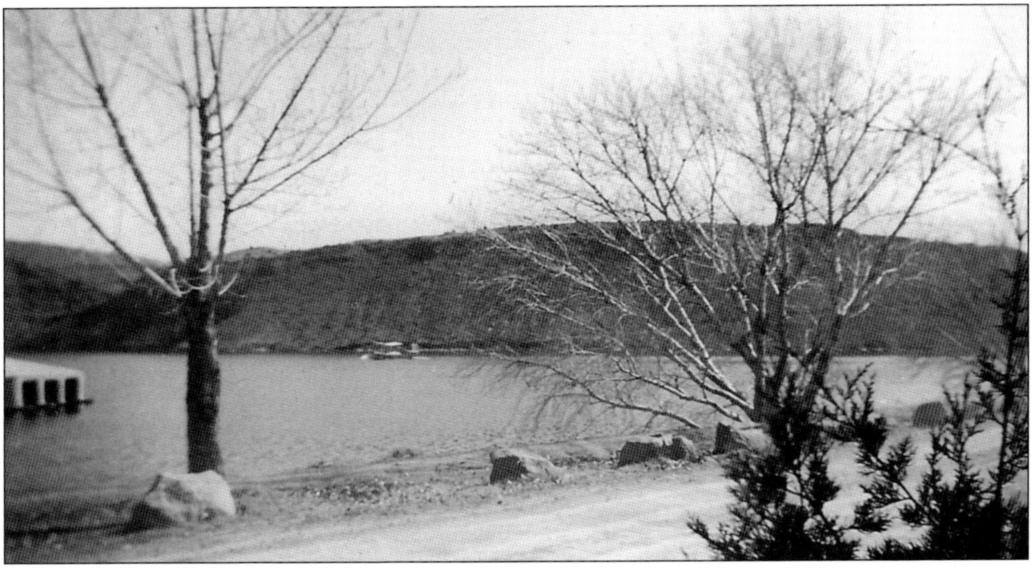

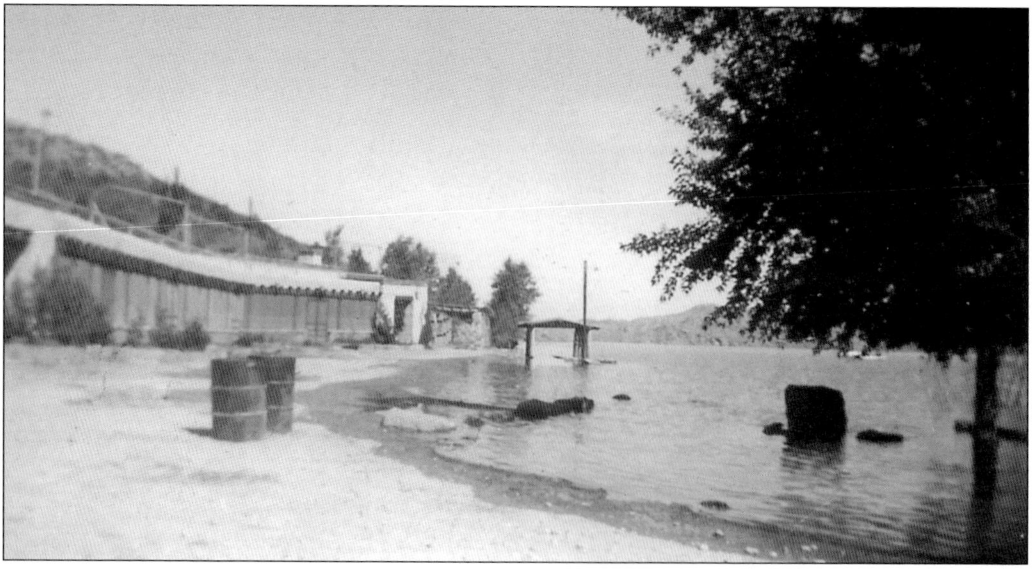

The *Las Cruces Sun News* reported on May 7, 1942, that the water level in Elephant Butte reservoir was "within two feet of the spillway crest, the highest point it has reached. L.R. Fiock, superintendent of the Rio Grande project, said the water might flow over the spillway crest within the next two weeks. For flood protective measures, water was being released at the rate of 6,000 cubic feet per second. The inflow was 3,500 cubic feet per second. Unless there are additional storms this month, Mr. Fiock said, there will be no increase in the volume of water released." Both of these photographs were taken during the high-water period.

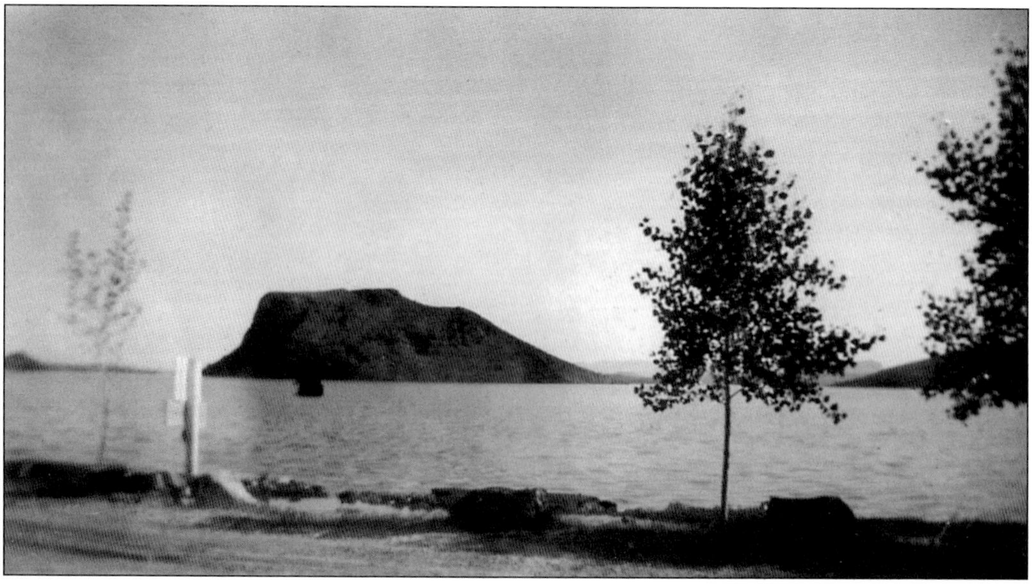

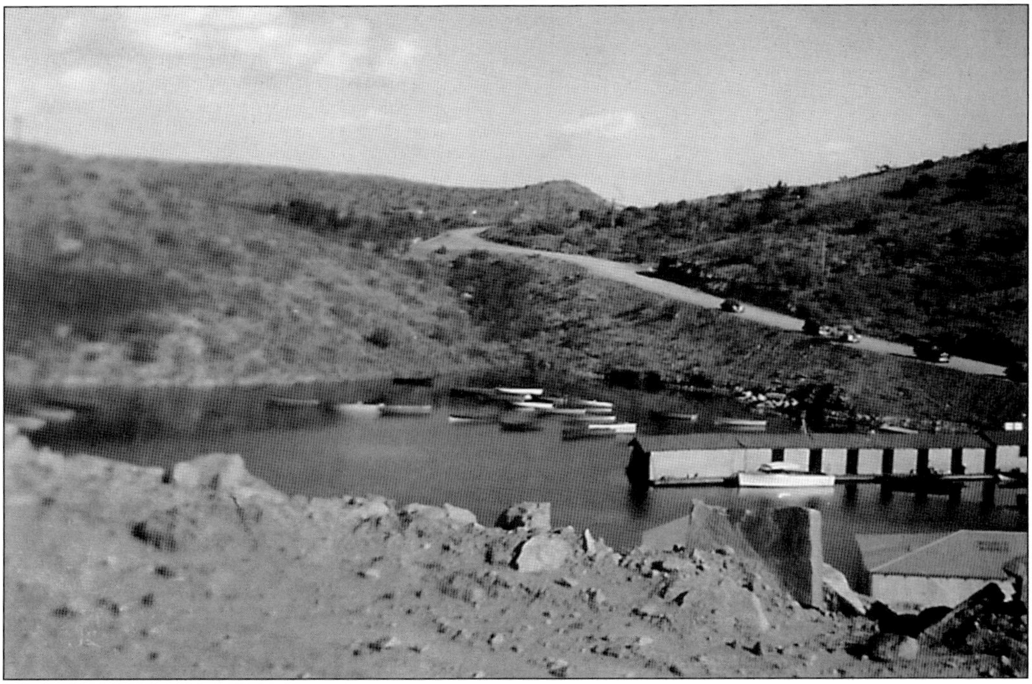

The *Albuquerque Journal* wrote on October 14, 1947, that "concessions at Elephant Butte Lake and Caballo reservoir are to be leased for a ten-year period instead of the former period of three to five years, in the hope that it will promote development of recreational facilities, according to L.R. Fiock superintendent of the Rio Grande project, who announces that bids for the concessions are to be opened at 2 p.m. on Nov. 1, in his office at the U.S. courthouse in El Paso. This new policy will be welcomed by the many New Mexico people who flock to Elephant Butte lake for fishing and boating and are interested in seeing the concession facilities developed."

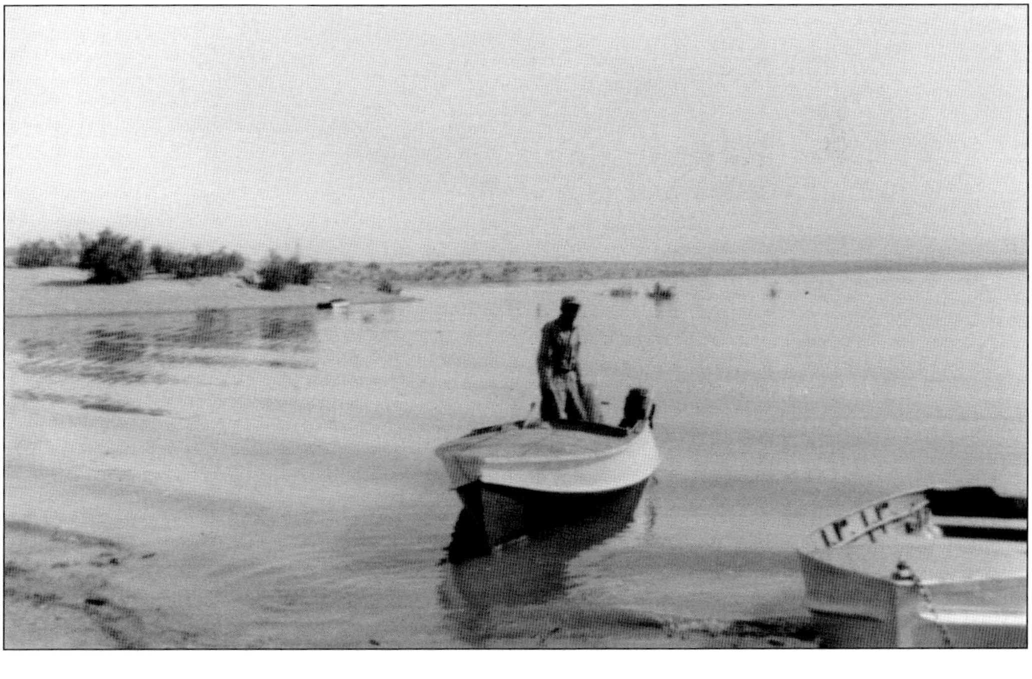

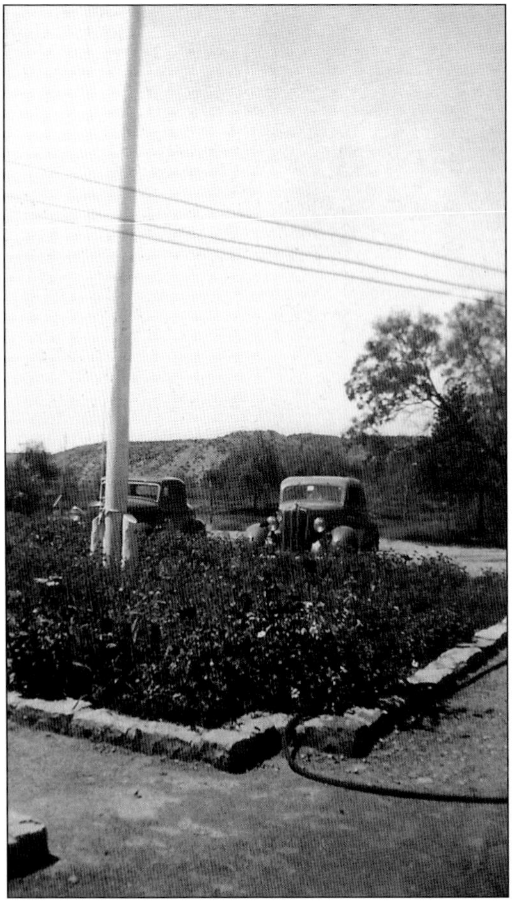

Pictured here are the beautiful grounds built by the CCC in the late 1930s at the fish hatcheries. Although it is no longer in operation, the hatcheries, under the supervision of Thomas F. Nelson of the US Fish and Wildlife Service, reared both trout and warm-water species that included largemouth bass, channel catfish, bluegill, sunfish, walleye, northern pike, white bass, crappie, and others. Brooder ponds were started on October 1, 1937, and would be completed two years later. They were a vital part of the hatchery, known as the Elephant Butte Fish Cultural Station, the Elephant Butte Fish Hatchery and most popularly, the Hot Springs National Fish Hatchery.

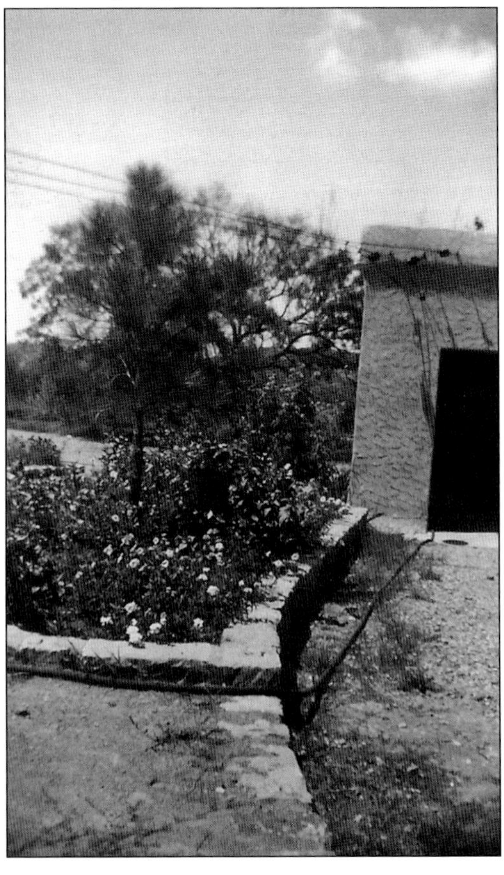

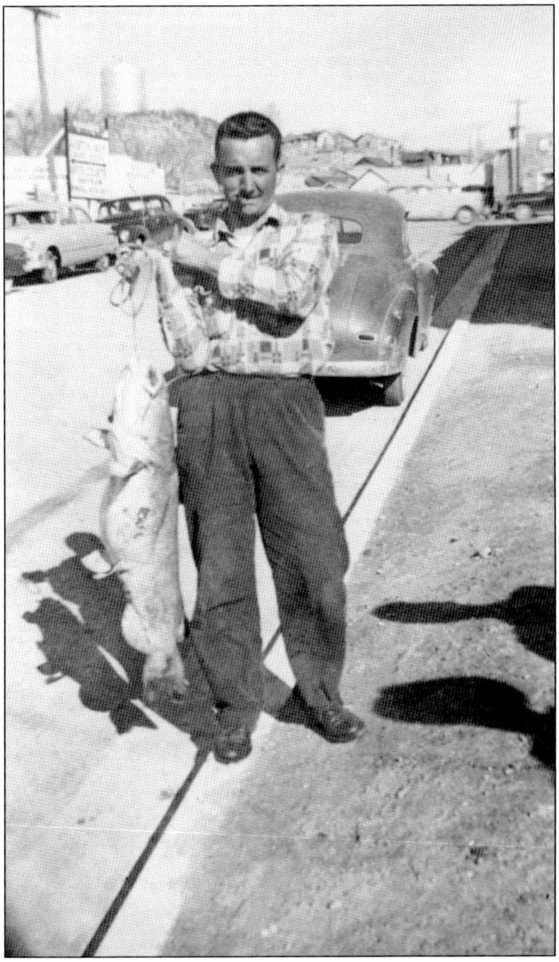

It was not uncommon for men and women to bring their catch down to the local bars or at least to the downtown area of Hot Springs, New Mexico (later renamed Truth or Consequences), for pictures and bragging rights. Fishing was big business and could also be profitable. In 1939, there were 25 bass tagged and released at Elephant Butte Lake. Each bass was worth a whopping $50 if caught. Glenn Mims, president of the Elephant Butte Hotel & Boat Company, had arranged for the stunt after the *Herald Post* in El Paso, Texas, had tagged fish at Cement Lake and Elephant Butte Lake. The two fish, nicknamed Popeye and Blackie, had been raised in the fish hatcheries in Elephant Butte and were worth $10 apiece. The man who caught Blackie could not bear to eat the beloved fish, so he had him mounted instead. No word on what happened to Popeye.

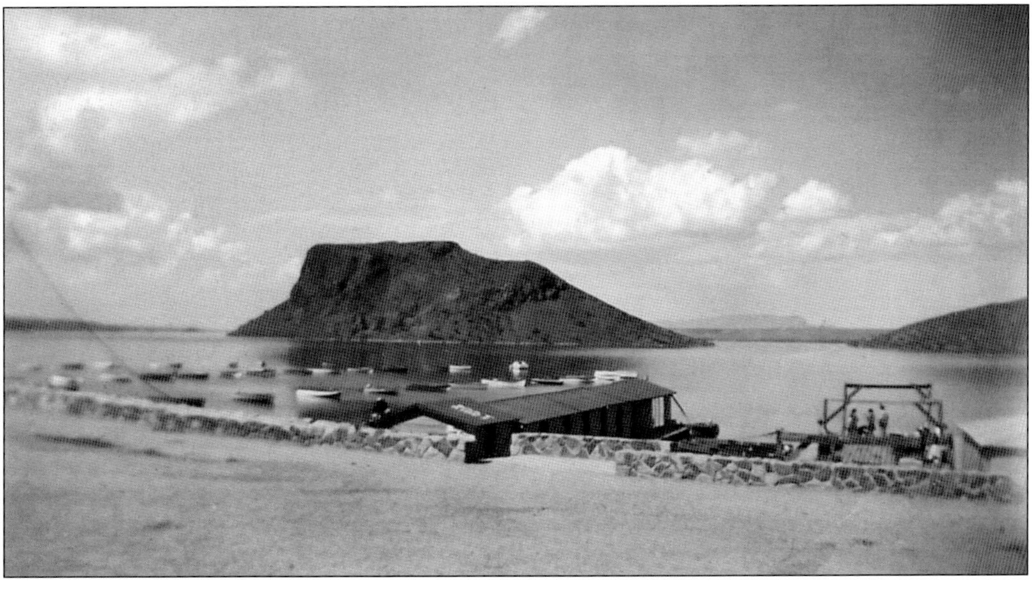

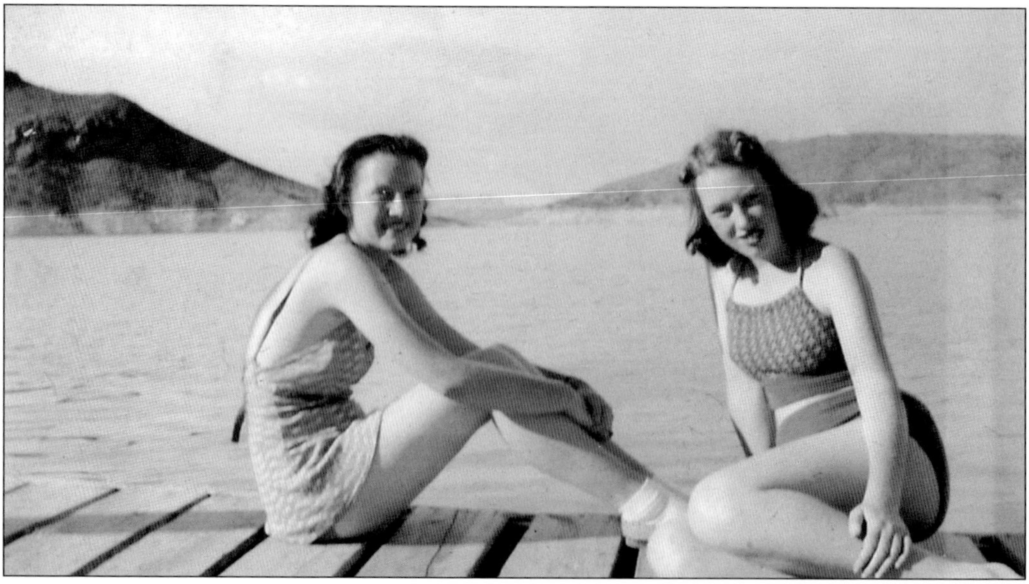

With bathing beauties and fast boats, the Hot Springs Regatta (located on Elephant Butte Lake) brought speedboat racers from all over the world to compete in boat races annually on the lake. There was said to be skiing and surfboarding to further wow the crowds. In June 1936, the fifth regatta was held at the lake, and C.E. Henson was the commodore for the event, which was held that year on June 22 and June 23. The honorary commodore was former New Mexico governor Clyde Tingley, who will be forever remembered for building the Carrie Tingley Crippled Children's Hospital in nearby Hot Springs, New Mexico. The hospital, named for Tingley's wife, Carrie, was built for children battling the devastating disease of polio.

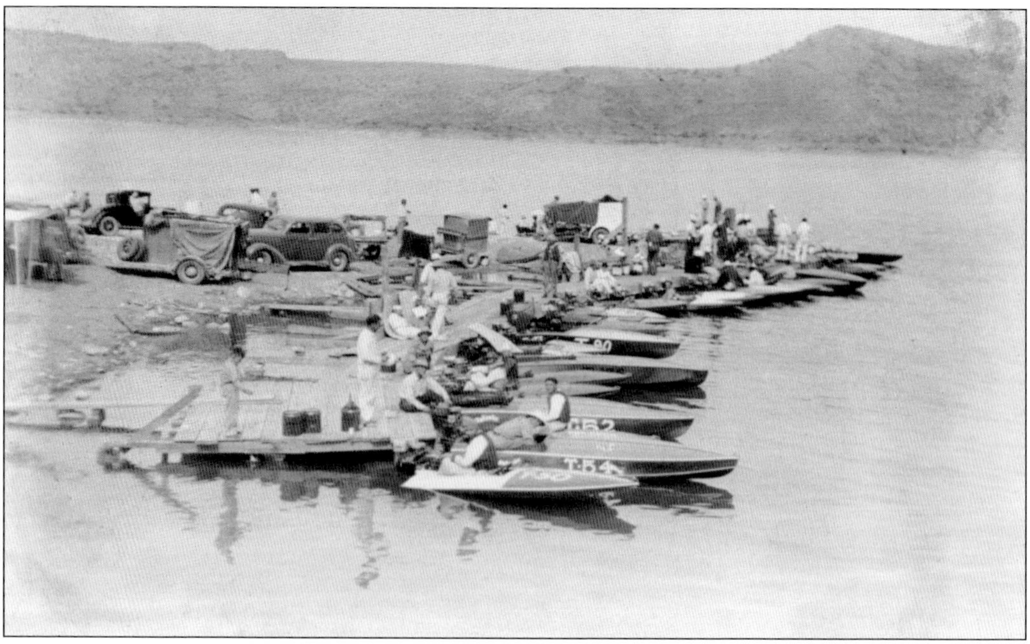

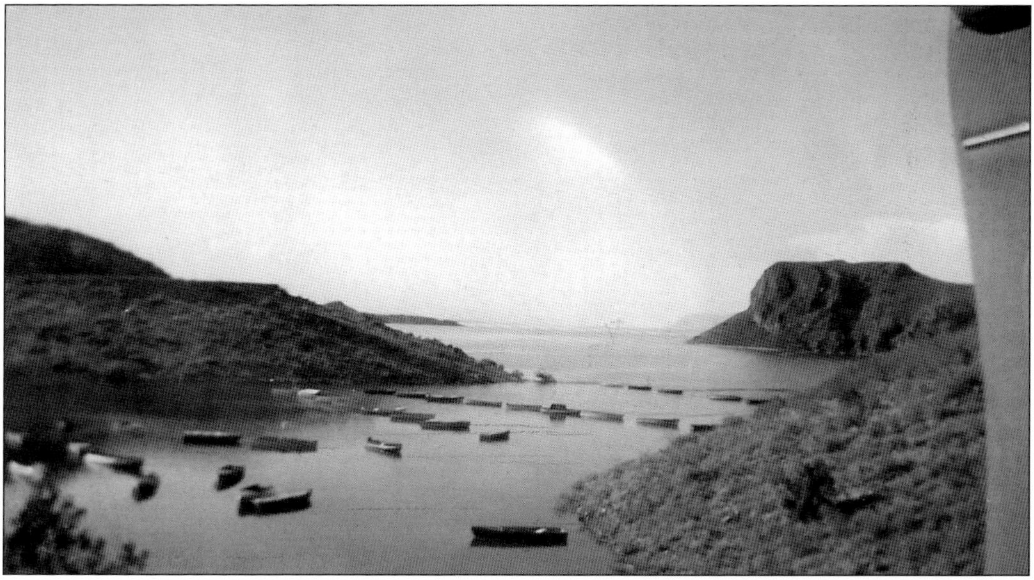

On May 12, 1940, the *Albuquerque Journal* printed an article by Thomas F. Nelson, foreman in charge. Nelson (later superintendent of the fish hatcheries) wrote, "Accommodations for week end parties, or parties for longer periods, are being prepared near the docks at the dam, Rock Canyon and The Narrows. These two and three-room cottages or cabins are being built by the Bureau of Reclamation and the Civilian Reclamation Corps. Several of the cabins are now finished, and others will be finished as rapidly as possible. It is certain the visitors or guests will be agreeably surprised, as these cabins will be finished as nice as the average home." Pictured on the right are Paul Woolford and his father in front of the hotel that had been built in 1936. (Both photographs courtesy of the Woolford family.)

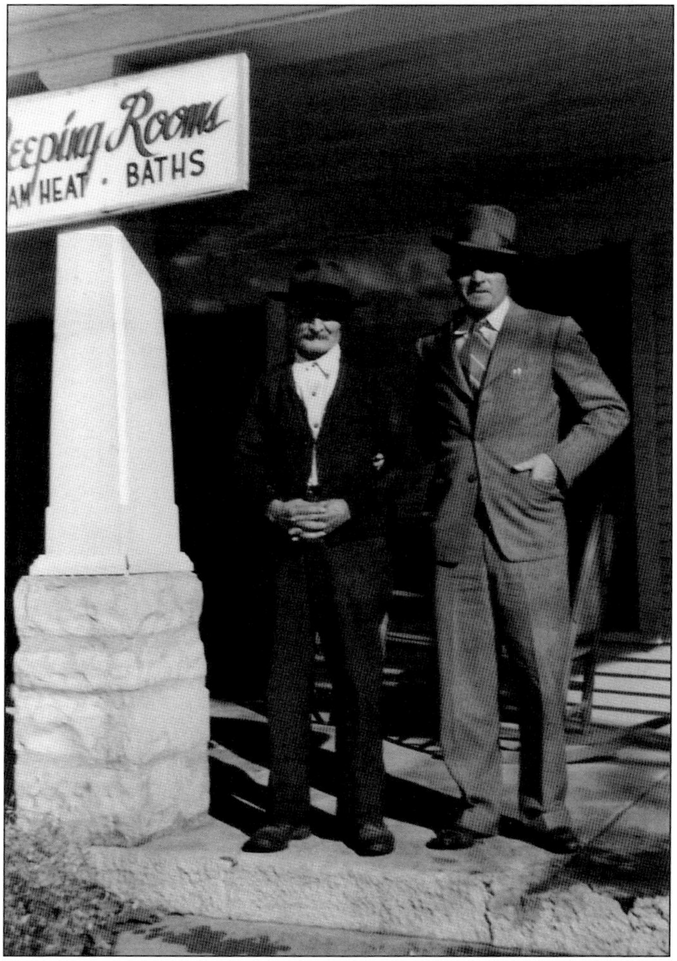

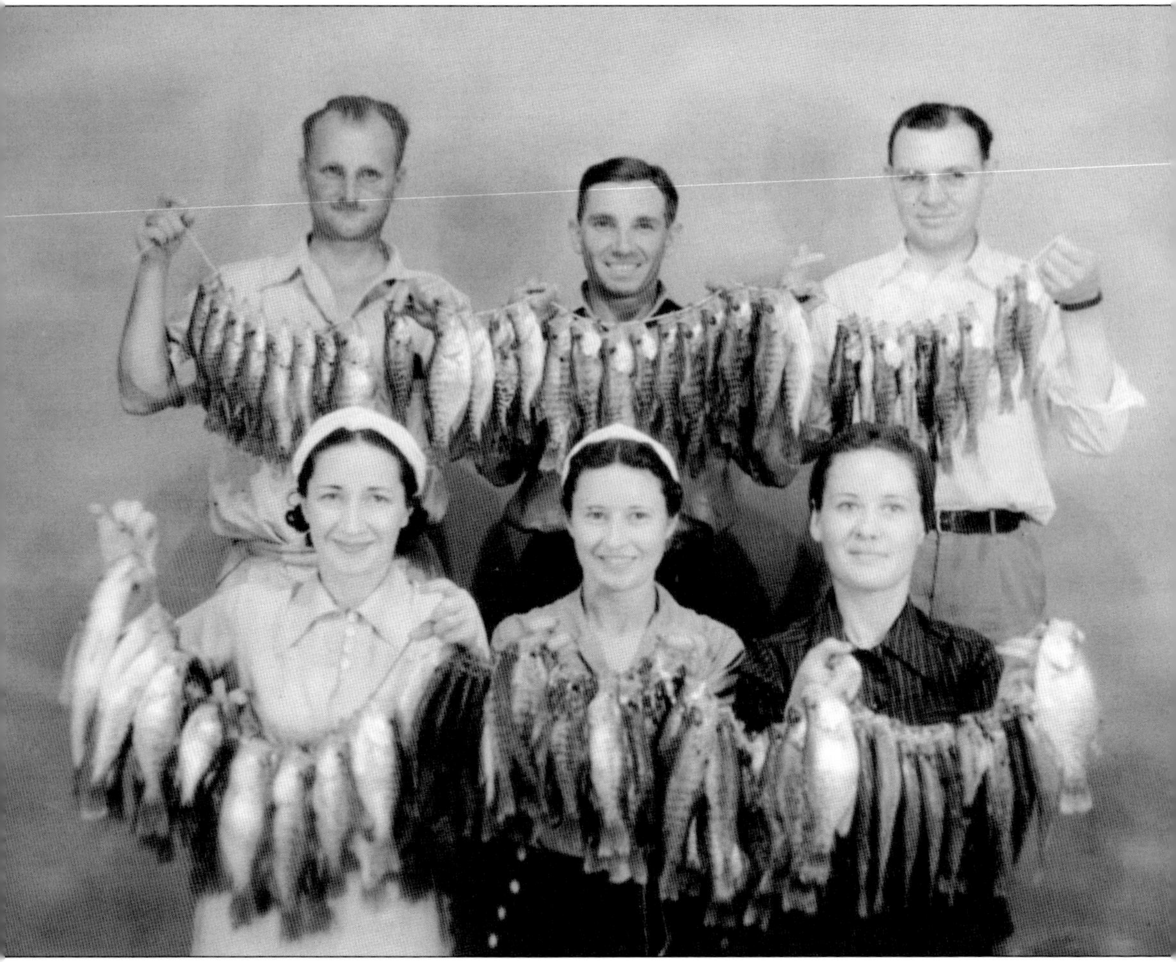

Fishing at Elephant Butte Lake was a sport that was not gender specific. Women not only fished, but they also clearly enjoyed the results, as noted by the big grins in the photograph. There was a day and time when fish caught at Elephant Butte Dam were paraded up and down the streets of Hot Springs, New Mexico, or brought into the local bars for a local fisherman (or woman) to exercise bragging rights. It was said that fishermen from Colorado, Kansas, Oklahoma, Texas, and Wyoming traveled to fish at Elephant Butte Lake. Striped bass, white bass, largemouth bass, crappie, walleye, and catfish are some of the various species popular at the lake. Various local newspapers bought the fishing news, as did major newspapers from Albuquerque, New Mexico. Fishing at the lake is still popular and, due to the fabulous weather, can be enjoyed year-round. (Courtesy of the Geronimo Springs Museum.)

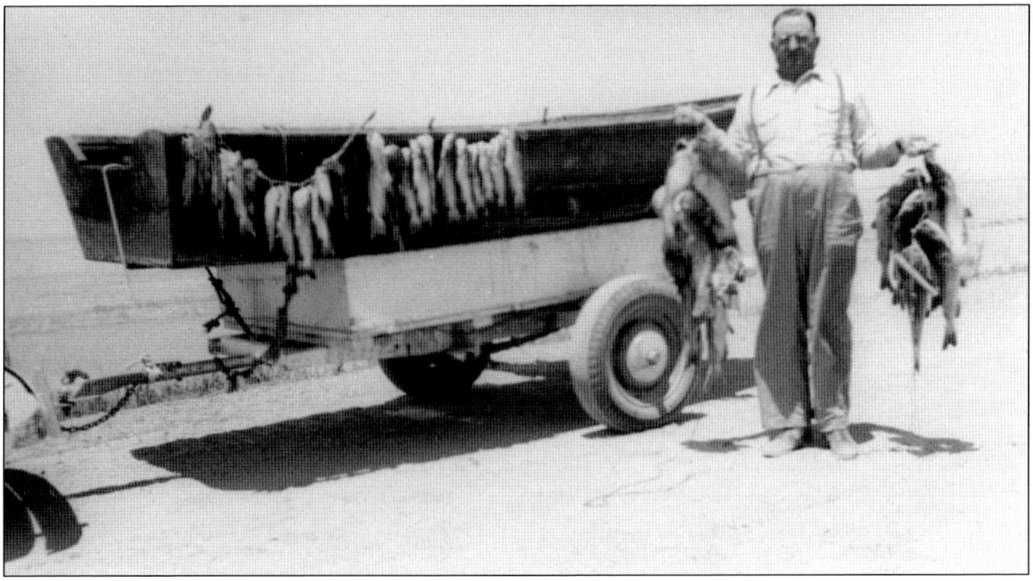

On May 12, 1940, the *Albuquerque Journal* reported, "Camping spaces have been provided at numerous places around the lake for parties who bring their own camp equipment. A large hotel is nearing completion and it is expected this will be ready for guests at an early date. Boats, motors, baits, etc. are available for hire, rent or sale at the docks near the dam, Rock Canyon and The Narrows. This concession is held by J.G. Mims, who has had this business for several years and is well known by the fishermen and who, with a competent crew of assistants, is able properly to take care of all parties. Mims has expanded considerably by building more and larger boat and fishing docks, by purchasing a good number of new Thompson boats and Johnson motors to sell or rent."

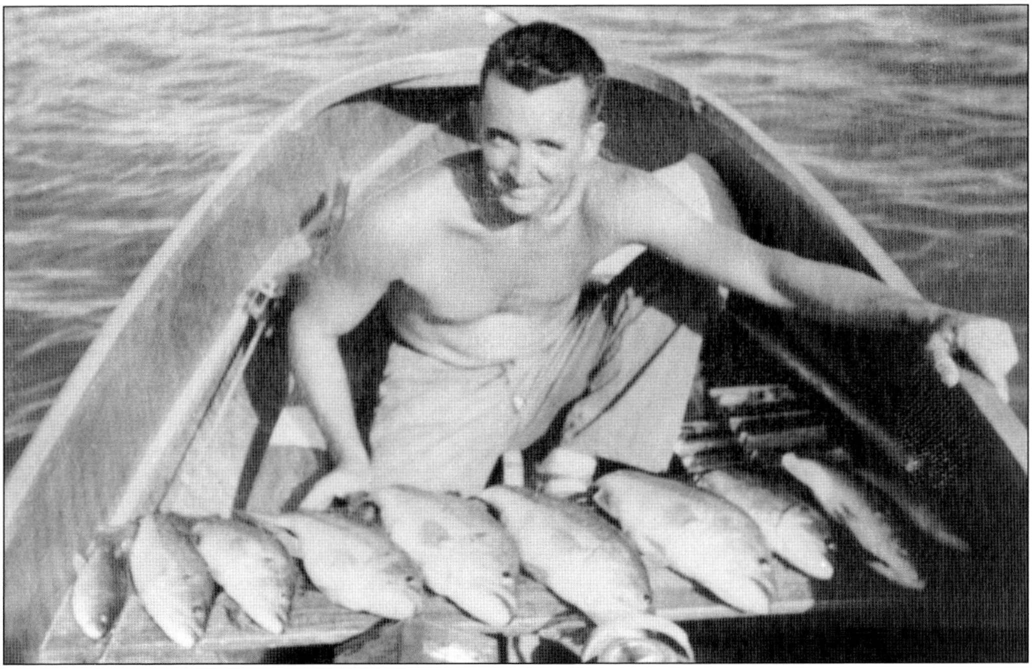

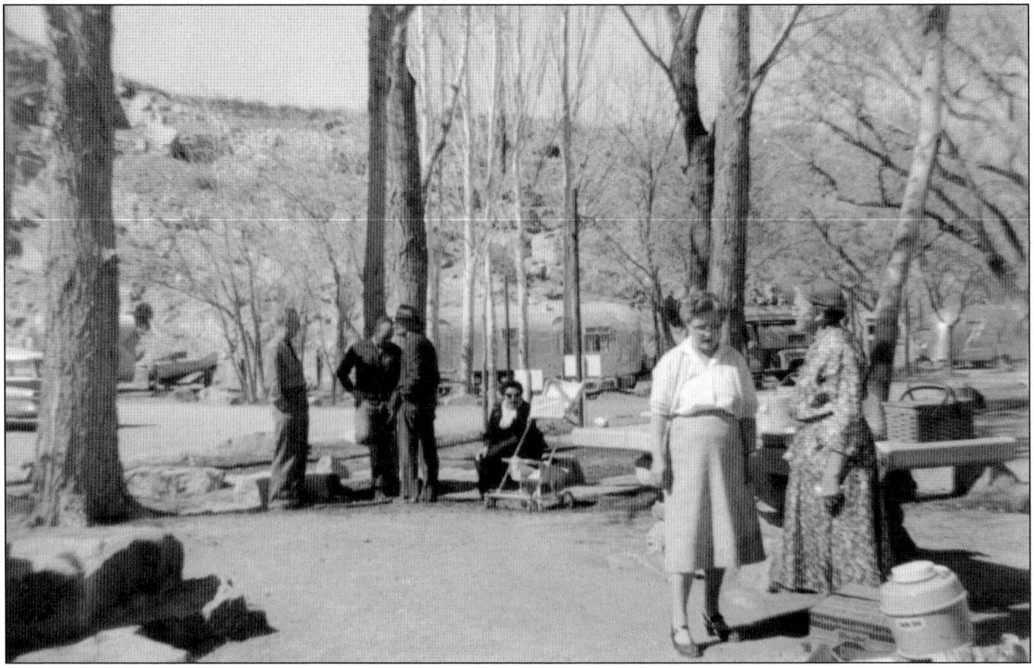

Both of these photographs are of the camping area behind the cottages at the damsite. During the 1940s, concessions included fast speed boats, a dining establishment by the dam, and many recreational areas for camping. According to the *Albuquerque Journal* of May 12, 1940, "This concession is owned and operated by Mr. and Mrs. Del Ritchie, who strive to please and guarantee satisfaction. In addition to and adjoining the café there is a large dance floor, cold drinks, groceries, lunch meats, fishing tackle, newspapers, etc, with no advance in prices." (Both photographs courtesy of the Geronimo Springs Museum.)

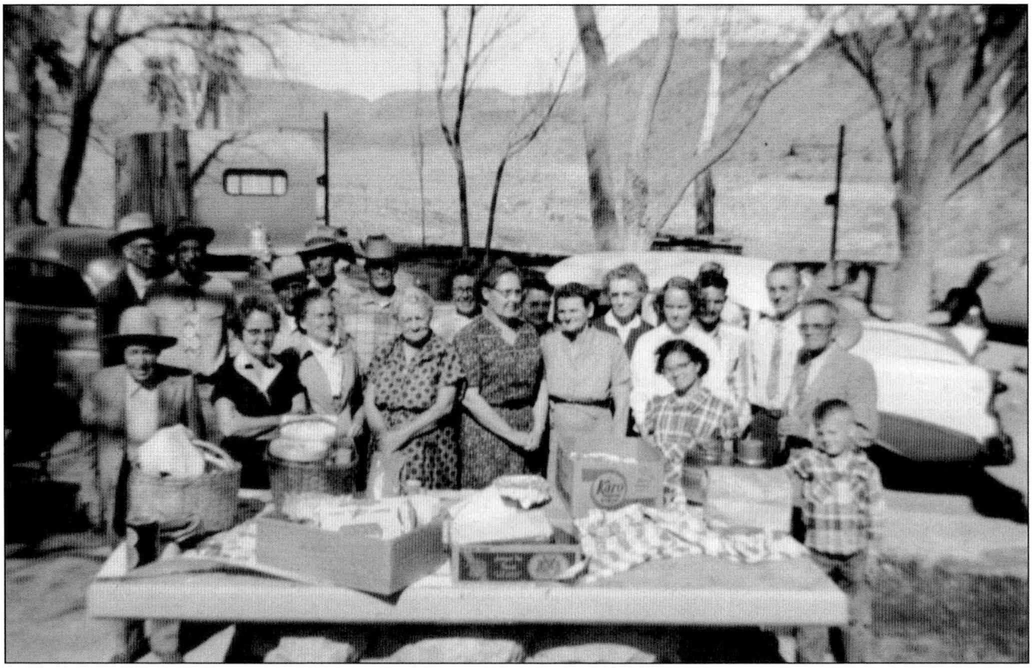

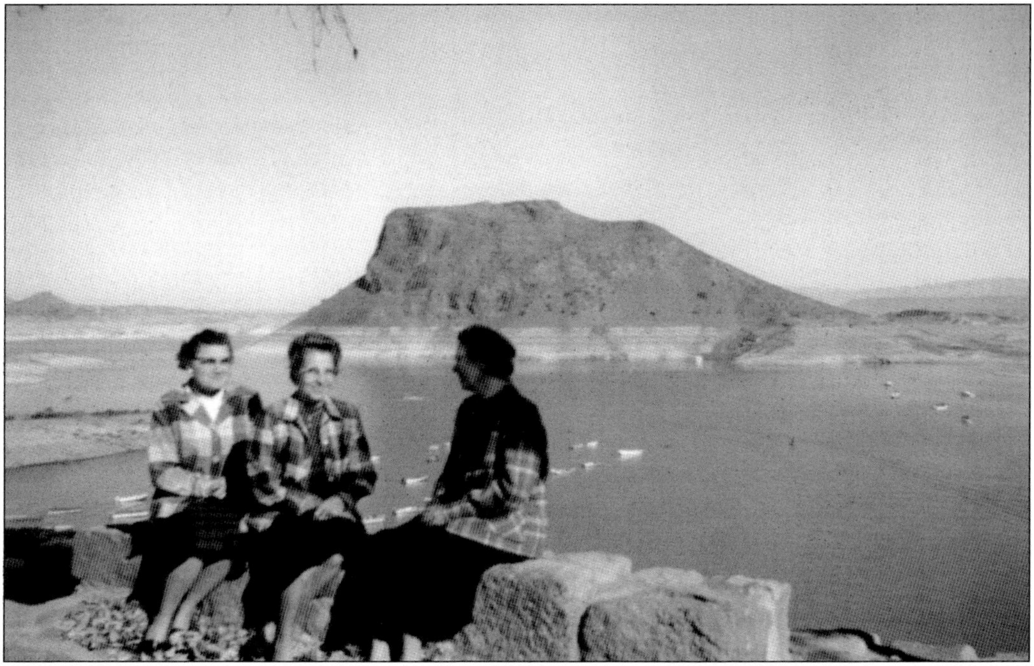

In order to provide access to the increasing number of people wanting to visit the lake and its amenities, the Bureau of Reclamation and the Civil Conservation Corps (CCC) built roads from the dam up the east shoreline of the lake. This connected the lake to the Hot Springs–Albuquerque Highway. The roads built allowed access to most points on the east shore from Elephant Butte Dam to the upper end of the lake. Game wardens from the New Mexico Game and Fish Departments and the US Biological Survey patrolled the lake. The *Albuquerque Journal* noted on May 12, 1940, that the "wardens are glad to be of assistance to strangers and to ensure you a pleasant and successful outing." (Both photographs courtesy of the Geronimo Springs Museum.)

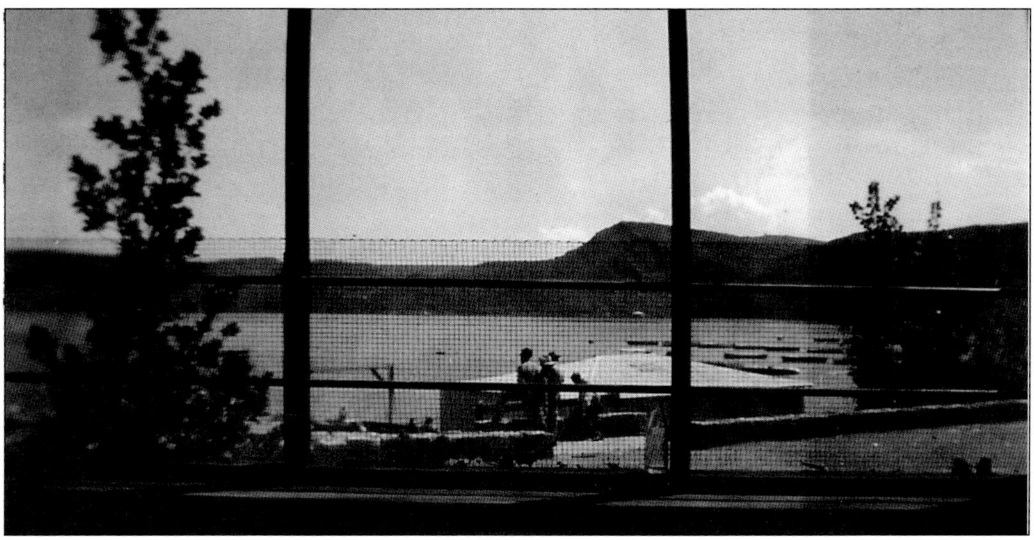

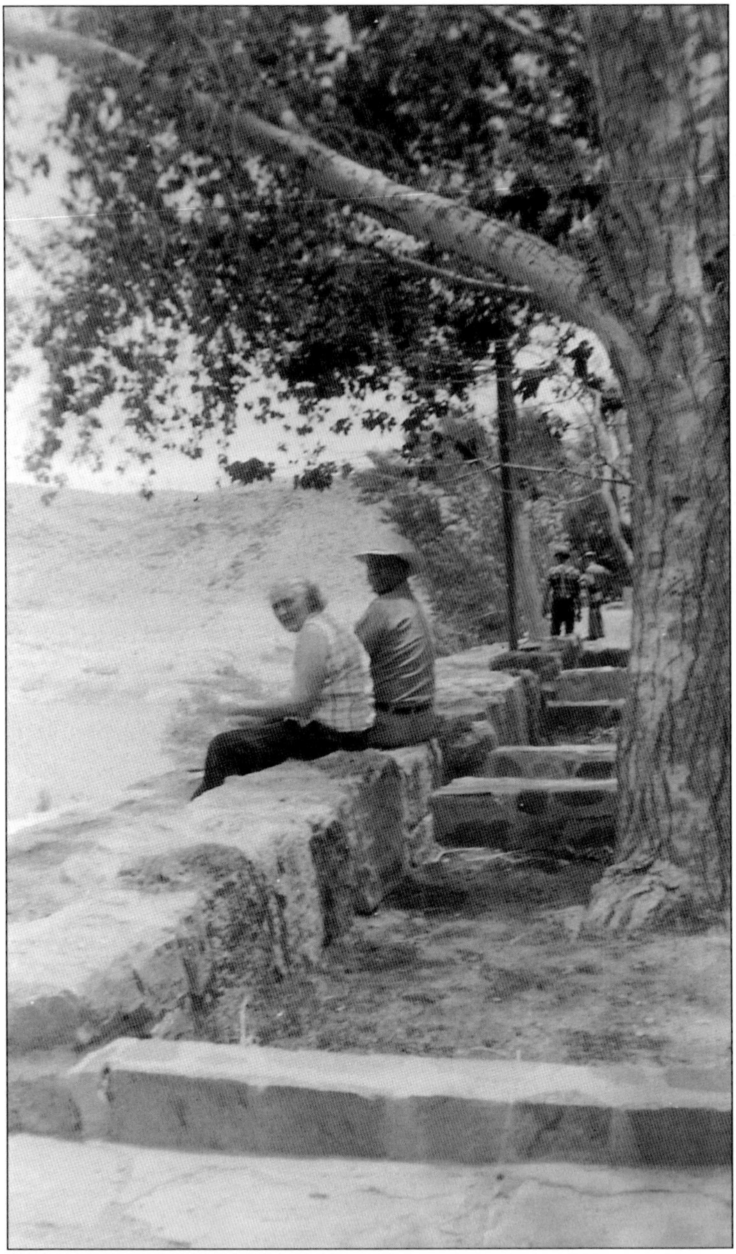

The National Park Service did a study in 1939 that consisted of questionnaires filled out by visitors to four different tourist areas in New Mexico: the Bottomless Lakes in Roswell, New Mexico; Hyde State Park near Santa Fe, New Mexico; Elephant Butte Lake; and the Tucumcari Metropolitan Park near Tucumcari, New Mexico. According to the results of the questionnaire, 92 percent of the individuals surveyed were in favor of visiting wide open spaces. Over 80 percent of the questionnaires filled out indicated that the head of the family had an annual income below $3,000, had an average of three weeks of vacation in a year, and enjoyed (in order of preference) fishing, swimming, and boating. Other activities included picnicking, touring, and camping. The couple above are sitting on the rock wall, an example of the stonework completed by the Civilian Conservation Corps (CCC).

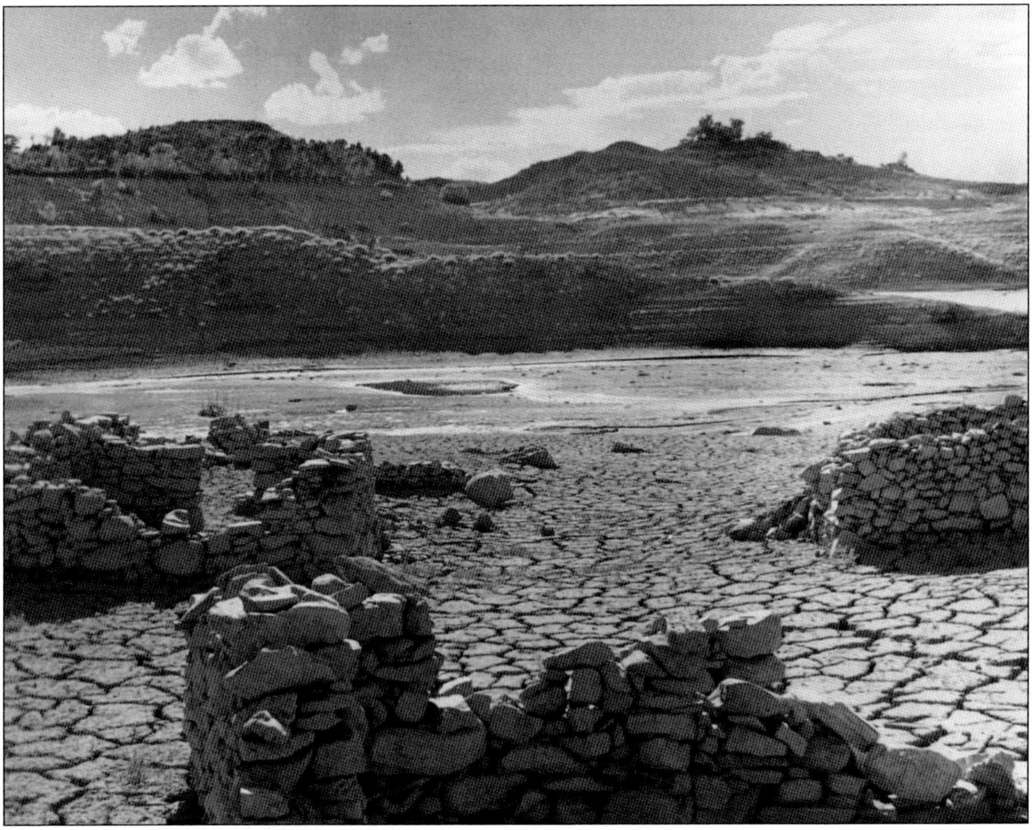

On August 6, 1954, Elephant Butte Reservoir was at 9,900 acre feet (elevation 4258.03 feet, 148.97 feet below the spillway crest). This was the lowest storage/elevation for Elephant Butte Reservoir for its period of record as of date (January 2015). It was at 12,000 acre feet in late July 1954, another extreme drought year.

Pictured here is the top of the butte at Elephant Butte Lake. Elephant Butte Lake State Park is considered to be in the Rio Grande rift, which is a northerly-trending continental rift that extends from Colorado to northern Mexico. This portion of New Mexico has been affected by compression, regional volcanism, and extension during the last 70 million years.

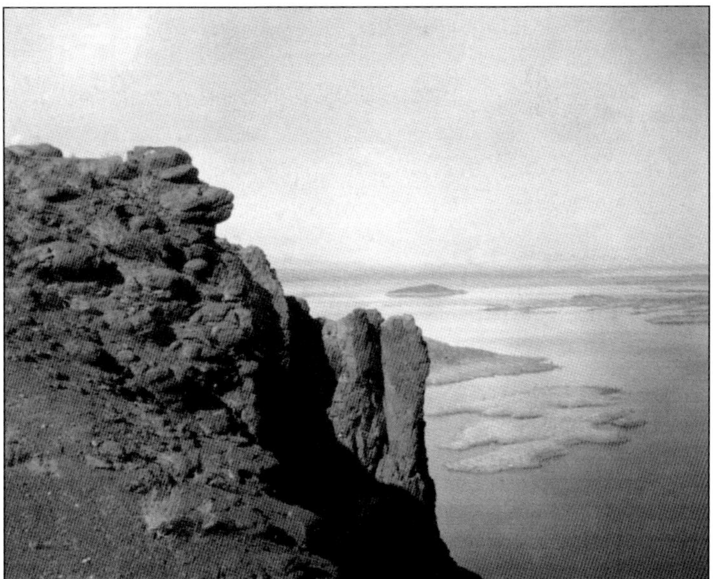

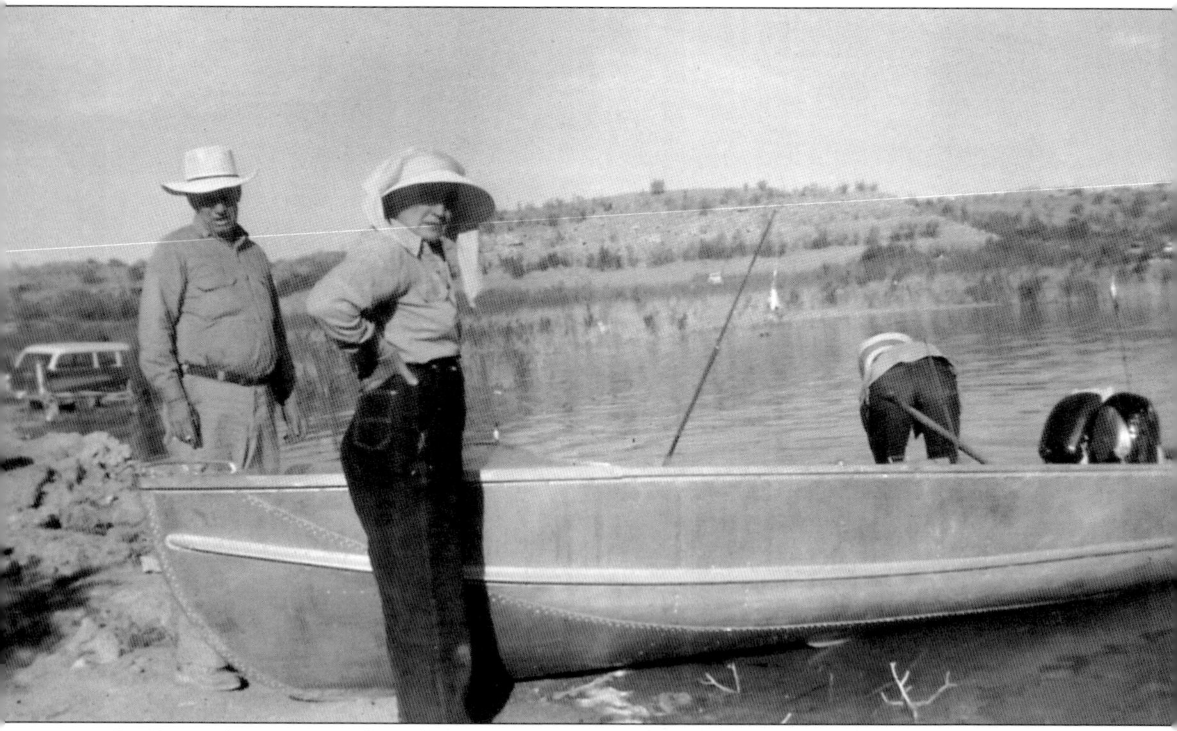

On December 7, 1951, the *Albuquerque Tribune* reported that Sen. Dennis Chavez wanted to change the name of Elephant Butte Lake to Martinez-Courchesne Lake. The Elephant Butte Water Rights Association and the Truth or Consequences Chamber of Commerce opposed the name change. Senator Chavez pointed out that the waters impounded by the dam had never been officially named, and he wanted to honor the pioneers of the lake. The article pointed out, "Of course, the opposition of the Truth or Consequences Chamber is somewhat ironical. That thriving little city near the lake changed its name from Hot Springs to the present unwieldy title two or three years ago. It did so in order to gain publicity from a radio program of that name." The senator was referring to the name of what was formerly Hot Springs—Truth or Consequences. The article went on to proclaim, "One thing is certain. Whether the lake continues to be called Elephant Butte Lake or gets a new name, the folks below the dam will continue to shout for more water from up the river."

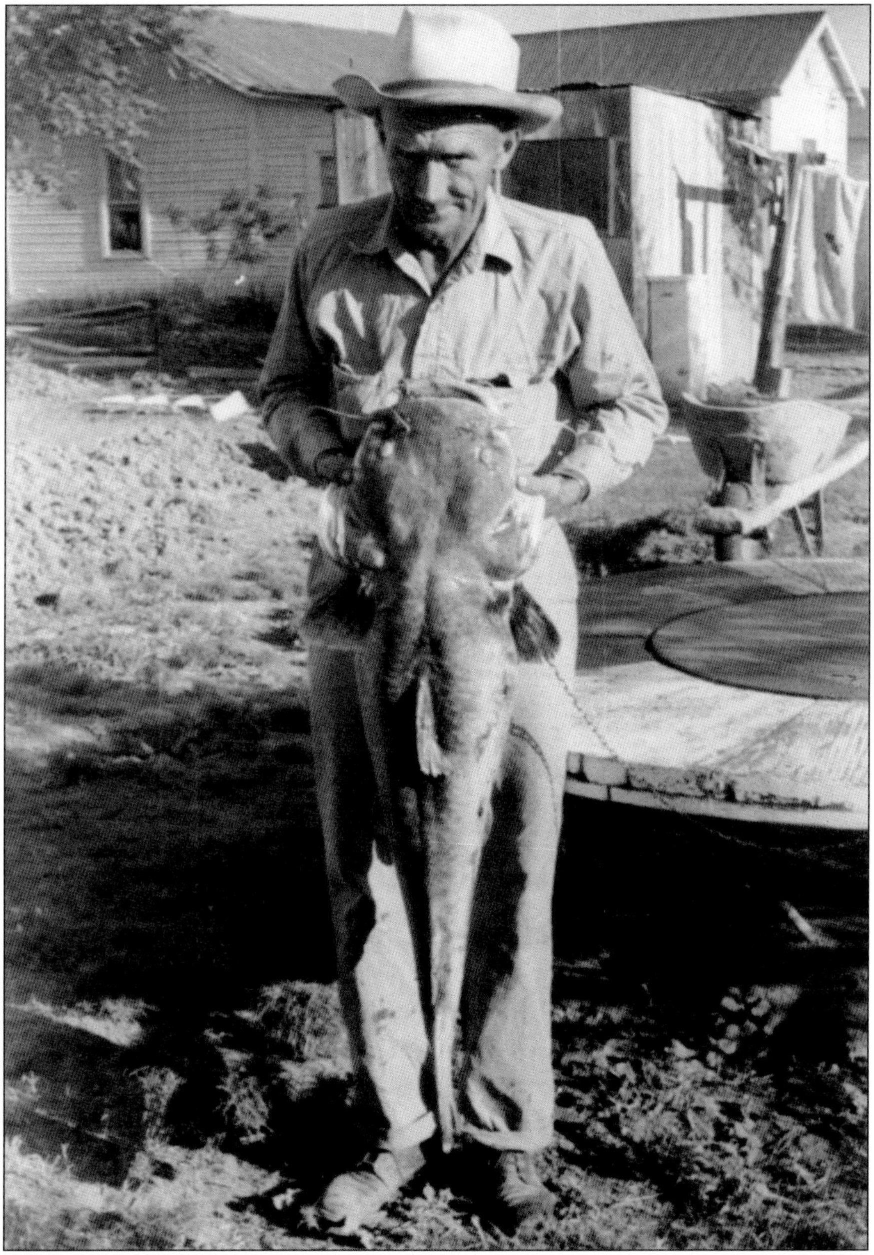

Tourist cabins were built in the area known as the Narrows in the late 1930s, and Horace Albert "Doc" Witt built one of the private cabins. In the photograph above, he is showing off his prize catch, a gigantic catfish. Witt was a former CCC camp worker from Las Cruces, and he owned Witt's Tackle and Marine there but spent every minute he could at Elephant Butte Lake. Doc, as he was known, carried the materials for his cabin, seen behind him, one truckload at a time from Las Cruces. He built the cabin with his own two hands. The Narrows was known as a favorite fishing spot for many a good fisherman. In 1939, the *Albuquerque Journal* printed an article on June 18 about a group of fisherman who caught, altogether, over 200 pounds of fish, with the largest fish topping the scale at 15 pounds. One of the fish caught in the Narrows was a prize blue channel catfish.

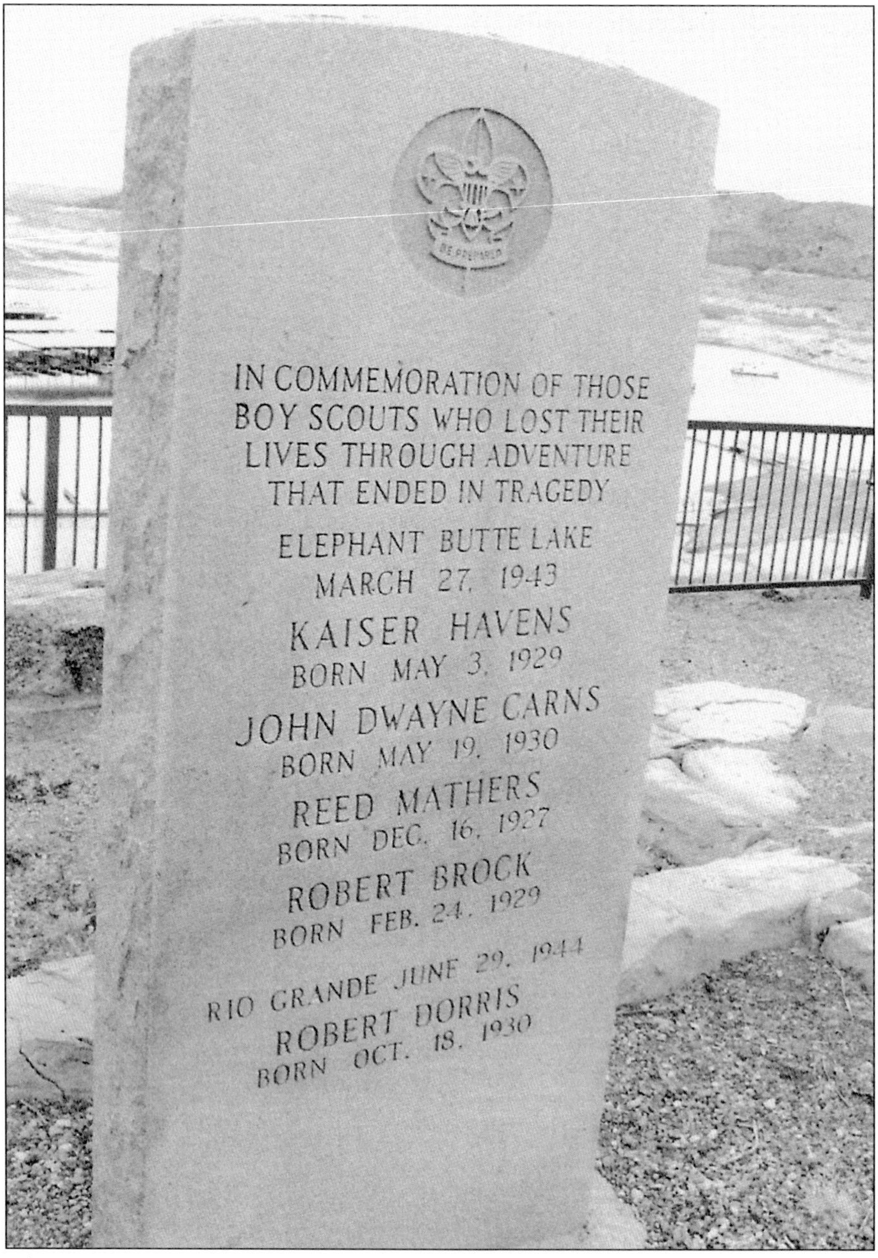

On March 29, 1943, four Hot Springs Boy Scouts drowned at Elephant Butte Lake. Only a fifth young man escaped drowning when the wind capsized the small boat the boys were using to ferry their supplies over to Rattlesnake Island. The boys ranged in age from 12 to 14 and were part of a camping party of "tenderfoot" Boy Scouts from Troop 70 at Hot Springs, New Mexico. The New Mexico State Police and Hot Springs residents supplied boats to aid in the search. Only one body would ever be recovered. On July 27, 1951, a monument made from South Carolina granite was furnished at a cost of $227.50 by the Maddox Monument Company of Roswell, New Mexico. It took T.B. Maddox, Carl Maddox, and Dr. E.R. Frost (chiropractor) two days to put the monument up. At the end of the list of names was the name of Robert Dorris, a young man who drowned in a separate accident in the Rio Grande.